Supply Chain Optimization under Uncertainty

Supply chain design for optimum performance

Barrie Michael Cole
BSc Chem. Eng.
MSc Eng.
Ph.D. Eng.

Vernon Press

www.vernonpress.com

In the Americas:
Vernon Press
1000 N West Street,
Suite 1200, Wilmington,
Delaware 19801
United States

In the rest of the world
Vernon Press
C/Sancti Espiritu 17,
Malaga, 29006
Spain

Library of Congress Control Number: 2014948701

ISBN: 978-1-62273-016-2

Foreword

This work discusses the planning and optimisation of the infrastructural configuration, i.e. the design, as well as the optimisation of operational performance of supply chains that are subject to conditions of **operational uncertainty**, i.e. fuzziness (<, ≤, >, ≥) and stochastic uncertainty that typically exist in a supply chain operational environment. The 'planning' aspect refers to an occasional need to accommodate multiple performance objectives in the assessment and management of supply chains, and this aspect is more commonly referred to as 'multi-objectivity, which means the existence of multiple maxima, multiple minima or a combination of both maxima and minima objectives in a supply chain environment.

This is an extension of prior thought on the subject where, previously, only one or two conditions of operational uncertainty were considered in a supply chain environment, and sometimes also the planning requirement, i.e. multi-objectivity, was involved. The discussions in this study then effectively represent an extension of previous thought on the subject in that by considering not only relevant cases of operational uncertainty, i.e. fuzziness and/or stochastic uncertainty, but also by considering those prevailing planning instances of multi-objectivity (i.e. maxima or minima or a combination of both maxima and minima) in a supply chain operational environment. Such capability would be tantamount to being able to deliver **'realistic'** and planned supply chain solutions since all prevailing conditions of operational uncertainty would have been accommodated. A typical supply chain is a production and distribution network consisting of multiple production centres, distribution facilities and sales outlets. The objective of this work is to introduce and define a methodology for the optimisation of supply chains under prevailing combinational conditions of uncertainty and planning, which would be tantamount to the means of finding the best operating solution for supply chains, Such methodology is formulated by identifying single (e.g. fuzzy optimisation), dual (e.g. fuzzy-multi-objective optimisation) and ternary (e.g. stochastic-fuzzy-multi-objective optimisation) instances of supply chain under uncertainty and planning techniques from previous research works, analysing them and then extracting the sequence of optimisation steps utilised. Such extracted optimisation techniques are then logically collated, according to established procedure, to create a methodology for the planning and optimisation of supply chains, under conditions of uncertainty.

This method is applied to the planning and optimisation of a NPK fertiliser production and distribution facility, which is subject to fuzzy market demand uncertainty and which also has a multi-objective operational planning requirement, to maximise the production and distribution of an entire range of NPK fertiliser in accordance with market demand, as well as to simultaneously minimise the generation and discharge of hazardous Hydrogen Fluoride (HF) gaseous effluent from the NPK fertiliser Nitrophosphate production unit. There are over 15 different blends of NPK (nitrogen, phosphorous, potassium) fertiliser available, with each blend being suited to a particular agricultural crop-type, e.g. maize, wheat, lucerne etc., and therefore the market demand uncertainty is directly translated into production uncertainty with uncertain raw material allocation in terms of the various sources of N, P and K, i.e. ammonium nitrate (NH_4NO_3), nitrophosphate ($(NH_4)_2SO_4$, $(NH_4)H_2PO_4$, NH_4NO_3, $CaSO_4.2H_2O$), superphosphate ($40\%Ca(H_2PO_4)_2 + 60\%CaSO_4.2H_2O$) and potassium chloride (KCl). Optimum production/distribution results revealed an achievement of 99.3% of maximum possible production and distribution capability, and also in accordance with market demand in conjunction with an effluent discharge, which was 94.1% of the minimum possible.

Acknowledgements

I would like to express my sincere thanks and appreciation to Prof Steven, Bradshaw of the University of Stellenbosch, SA, and to Prof Herman Potgieter, of the University of Manchester, UK, for their contributions into the development of this concept.

I would also like to express my special gratitude and thanks to the Omnia Group of South Africa, and especially to certain staff members, for permitting and assisting me in basing my supply chain application example on the production and distribution operations of their NPK fertiliser supply chain.

Table of Contents

List of Tables

List of Figures

Chapter 1

INTRODUCTION

1.1 Purpose

The primary intention of this work is to provide guidelines to the various custodians of supply chain management, be they operational or management in obligation, so that optimal designs as well as optimal performances may be achieved for those operationally ***uncertain*** supply chains that they are responsible for. For the sake of clarity, it must be borne in mind that optimal supply chain design normally implies lowest possible supply chain infrastructure cost whereas optimal supply chain performance normally equates to highest possible supply chain throughput or profitability. A supply chain infrastructure is typically defined as being comprised of an integrated network of permissibly interconnected supply chain units, i.e. production or manufacturing plants with associated inventory management capability, distribution centres or warehouses and retail sales outlets. Moreover, supply chain uncertainty is defined as the relevant combination of fuzzy uncertainty or imprecision, e.g. product demand is in the region of 2,500 t/mth, and/or stochastic, or probabilistic, uncertainty that exists in a supply chain environment whereas supply chain planning refers to the number and type of operational objectives to be achieved.

Therefore, the primary purpose of this work is to: *define and derive a planning, i.e. multi-objective, and optimisation methodology for a supply chain network that is operating under conditions of uncertainty, i.e. fuzzy and stochastic uncertainty, and that can be successfully applied to any Supply Chain Network*

1.2 Key Definitions

Supply Chain Management

Interest in **supply chain management** has steadily increased since the 1980s when firms saw the benefits of collaborative relationships within and beyond their own organisation. Firms are finding that they can no longer compete effectively in isolation of their suppliers or other entities in the supply chain. Interest in the concept of supply chain management has steadily increased since the 1980s when companies saw the benefits of collaborative relationships within and beyond their own organisation. A number of definitions have been proposed concerning the concept of "the supply chain" and its management. The term, supply chain management, does not replace supplier partnerships nor is it a description of the logistics function. Industry groups are now working together to improve the integrative processes of supply chain management and accelerate the benefits available through successful implementation. The competitive importance of linking a firm's supply chain strategy to its overall business strategy and some practical guidelines are offered for successful supply chain management.

Definitions of the term, 'supply chain', abound and since a number of definitions of supply chain management have been proposed in the literature and in practice, this work defines the concept of supply chain management and discusses its historical evolution. The term does not replace supplier partnerships, nor is it a description of the logistics function. The competitive importance of linking a firm's supply chain strategy to its overall business strategy and some practical guidelines are offered for successful supply chain management. Various definitions of a supply chain have been offered in the past several years as the concept has gained popularity.

The *APICS Dictionary* describes the supply chain as:

1. The processes from the initial raw materials to the ultimate consumption of the finished product linking across supplier-user companies; and
2. The functions within and without a company that enable the value chain to make products and provide services to the customer.

Another source defines supply chain as, the network of entities through which material flows. Those entities may include suppliers, carriers, manufacturing sites, distribution centres, retailers, and customers. The Supply Chain Council (1997) uses the definition: "The supply chain - a term increasingly used by logistics professionals - encompasses every effort involved in producing and delivering a final product, from the supplier's supplier to the customer's customer. Four basic processes - plan, source, make, deliver - broadly define these efforts, which include managing supply and demand, sourcing raw materials and parts, manufacturing and assembly, warehousing and inventory tracking, order entry and order management, distribution across all channels, and delivery to the customer." Occasionally, a supply chain is defined as "all of those activities associated with moving goods from the raw-materials stage through to the end user. This includes sourcing and procurement, production scheduling, order processing, inventory management, transportation, warehousing, and customer service. Importantly, it also embodies the information systems so necessary to monitor all of those activities."

Further to the definition of a supply chain, the concept of supply chain management has also been defined as, “An integrating philosophy to manage the total flow of a distribution channel from supplier to ultimate customer". It has also been stated that integrated supply chain management’ is all about “migrating from the external customer and then managing the processes that are needed to provide that customer with value in a horizontal way". It is believed that it is supply chains, and not firms, that compete and that the strongest competitors will be those that can provide management and leadership to the fully integrated supply chain solution, including external customers as well as prime suppliers, their suppliers, and their suppliers' suppliers".

From these definitions, a summary definition of the supply chain management may be stated as;

All the activities involved in delivering a product from raw material through to the customer including sourcing raw materials and parts, manufacturing and assembly, warehousing and inventory tracking, order entry and order management, distribution across all channels, delivery to the customer, and the information systems necessary to monitor all of these activities.

Supply chain management therefore coordinates and integrates all of these activities into a seamless process. It links all of the partners in the chain, including departments within an organisation, and all the external partners, including suppliers, carriers, third-party companies, and information systems providers. Managers in companies across the supply chain take an interest in the success of other companies. They work together to make the whole supply chain competitive. They have the facts about the market, they know a lot about competition, and they coordinate their activities with those of their trading partners. Supply chain management encompasses the processes necessary to create, source, to make, and to deliver demand. It uses technology to gather information on market demands and exchange information between organizations. A key point in supply chain management is that the entire process must be viewed as one system. Any inefficiencies incurred across the supply chain (suppliers, manufacturing plants, warehouses, customers, etc.) must be assessed to determine the true capabilities of the process within the supply chain.

Logistics management

Logistics management is the management of the flow of goods between the point of origin and the point of consumption in order to meet some requirements, for example, of customers or corporations. The resources managed in logistics can include physical items, such as food, materials, animals, equipment and liquids, as well as abstract items, such as time, information, particles, and energy. The logistics of physical items usually involves the integration of information flow, material handling, production, packaging, inventory, transportation, warehousing, and often security. The complexity of logistics can be modeled, analysed, visualised, and optimised by dedicated simulation software. The minimisation of the use of resources is a common motivation in logistics for import and export.

1.3 History of Supply Chains

The history of the supply chain initiative can be traced to early beginnings in the textile industry with the quick response program and later to efficient consumer response in the grocery industry. More recently a variety of companies across many industries have begun looking at the entire supply chain process. This section will discuss those early beginnings of the supply chain, the fundamental rationale for them and some of the more recent success stories.

Quick response, for general merchandise retailers and their suppliers

As a result of intense competition in the textile and apparel industry world-wide, leaders in the US apparel industry formed the Crafted With Pride in the USA Council in 1984 (Kurt Salmon Associates, Inc., 1993). In 1985, Kurt Salmon Associates were commissioned to conduct a supply chain analysis. The results of the study showed the delivery time for the apparel supply chain, from raw material to consumer, was 66 weeks long, 40 weeks of which were spent in warehouses or in transit. The long supply chain resulted in major losses to the industry due to financing the inventory and lack of the right product in the right place at the right time.

The result of this study was the development of the quick response (QR) strategy. QR is a partnership where retailers and suppliers work together to respond more quickly to consumer needs by sharing information. Significant changes as a result of the study were the industry adoption of the UPC code used by the grocery industry and a set of standards for electronic data interchange (EDI) between companies. Retailers began installing point of sale (POS) scanning systems to transfer sales information rapidly to distributors and manufacturers. "QR maximizes the profitability of inventory by placing the company's dollars where and when they are needed based on point of sale data plus sales history" (Mullin, 1994). QR incorporates marketing information on promotion, discounts, and forecasts into the manufacturing and distribution plan.

Efficient consumer response, the grocery business initiative

In 1992, a group of grocery industry leaders created a joint industry task force called the efficient consumer response (ECR) working group. The group was charged with examining the grocery supply chain to identify opportunities to make the supply chain more competitive (Kurt Salmon Associates Inc., 1993). Kurt Salmon Associates were engaged by the group to examine the grocery supplier/ distributor/consumer value-chain and determine what improvements in cost and service could be accomplished through changes in technology and business practices.

The results of the study indicated little change in technology was required to improve performance, other than further development of EDI and POS systems. However, the study identified a set of best practices which, if implemented, could substantially improve overall performance of the supply chain. As Kurt Salmon and Associates (1993) had found: "By expediting the quick and accurate flow of information up the supply chain, ECR enables distributors and suppliers to anticipate future demand far more accurately than the current system allows". Through implementation of best practices they projected an overall reduction in supply chain inventory of 37 per cent, and overall cost reductions in the industry in the range of $24 to $30 billion.

The successful adoption of ECR for a manufacturer depends on their ability to maintain manufacturing flexibility, which would enable them to match supply with demand. Key to this flexibility is a process that tightly integrates demand management, production scheduling, and inventory deployment to allow the company to better utilize information, production resources, and inventory (Weeks and Crawford, 1994).

A further development of ECR was the concept of continuous replenishment (CRP). CRP is a move away from pushing product from inventory holding areas to pulling products onto grocery shelves based on consumer demand (ECR Performance Measures Operating Committee, 1994). Point of purchase transactions are forwarded by computer to the manufacturer allowing them to keep the retailer replenished and balanced just-in-time. CRP has been introduced by a number of manufacturers (Garry, 1994). Procter & Gamble and Campbell soup are delivering as much as 30 to 40 per cent of their volume by CRP. Ralston, General Mills and Pillsbury distribute about 10 per cent by CRP. Estimates of improvements in performance with CRP include increasing inventory turns from 10 up to 50, reducing days of supply from 30 to 5 and increasing net margin from 5 per cent to 7 per cent.

1.4 Early supply chain initiatives

Apart from the apparel and grocery industry initiatives, other early manufacturing efforts to improve supply chain performance have been documented. Some of these include: Hewlett-Packard, Whirlpool, Wal-Mart, West Co., Becton Dickinson, Baxter, and Georgia-Pacific Corp. A brief outline of their various supply chain initiatives is described:

Hewlett-Packard

The computer components manufacturer systematically linked its distribution activities with its manufacturing activities in the computer terminal business in the early 1990s (Hammell and Kopczak, 1993). The implementation included changes in both the physical distribution of the product, and a new distribution requirements planning (DRP) system. The DRP system nets customer orders with forecasts and serves as the beginning pull in the supply chain.

Whirlpool

This appliance manufacturer commenced its supply chain performance improvement initiatives with a team of executives in 1992 chartering the vision - "Winning companies in the future will be those that come closest to achieving an inter-enterprise pull system. They will be linked in a short cycle response mode to the customer" (Davis, 1995). Whirlpool had created a new vice-president of logistics position who had established cross-functional teams for key product areas that had entered into single source agreements with certain suppliers. Such agreements were built around product reliability and were therefore utilised in product design. EDI (Electronic Data Interchange) was utilised to communicate with suppliers on a regular basis, and which was integral to the supply chain management program. As a result of all this, product availability was increased to the 90-95 per cent range, inventories levels were reduced by 15 - 20 per cent and product lead times were reduced as well.

Wal-Mart

This company commenced its supply chain initiative by working directly with certain key manufacturers (Johnson and Davis, 1995). The manufacturers were responsible for managing Wal-Mart's warehouse inventory levels and, as a result, Wal-Mart demanded and achieved a near 100 per cent order fulfilment rate on those products. KMart and other large retailers had implemented similar VMI (Vendor Management Inventory) programs.

West Co., Becton Dickinson, and Baxter

Within the medical products industry, three firms engaged in mutual supply chain relationships in the early 1990s (Battagia, 1994). Working together, at all management levels, all three companies had achieved significant improvements in quality and service, whilst, simultaneously, reducing their cycle times and costs.

Georgia-Pacific Corp.

A leader in the manufacturing and distribution of building products in North America, Georgia-Pacific began implementing supply chain management practices within the decentralised operations of their company (Blackwell, 1994). Previously, traffic managers in each division had controlled inbound and outbound shipments for each division. Shipping priorities were fragmented and internal and external customers were not satisfied. A new centralised Transportation and Logistics Division was created to coordinate and streamline the distribution process. The new division looked at needs and priorities across the business units and had realised savings to the company in terms of reduced freight costs and other logistics improvements.

Many other examples of companies successfully implementing their respective supply chain strategies are available (Blaser and Westbrook, 1995; Cook and Rogowski, 1996; Semich, 1994). The high level of interest in this topic clearly indicates that supply chain management had become a key strategic issue amongst a diverse group of companies, which resulted in steps being continually taken to improve customer delivery whilst, simultaneously, reducing overall costs.

1.5 Growth of interest in supply chains in the 20th century

Why had the effective management of supply chains become such a key issue during the 1990s? In part, the answer lies in the fact that very few of these supply chain companies continued to operate as vertically integrated entities. Companies had become far more specialised, which resulted in the search for suppliers who could provide certain quality materials at low cost, rather than them providing them themselves. It had become critical for such companies to manage the entire network of supply, themselves, in order for them to achieve optimum performance. They had realised that whenever a company dealt with another company, as part of the supply chain, both party's stood to benefit from the other party's success.

A second reason for increased supply chain interest partially stemmed from increased national and international competition. Multiple sources had become available from which customer demand could be satisfied and also alternative distribution channels had become available, from which customer accessibility could be maximised, and also at the lowest possible cost. Prior to this, companies had solved the distribution problem through the provision and maintenance of inventory levels at different locations throughout the entire supply chain. However, this had not been very economical since the dynamic nature of the marketplace had necessitated the maintenance of reasonable inventory levels, which was both risky and unprofitable. Further, the changing buying habits of customers taken together with the changing nature of products contribute greatly to this unprofitable situation. Such situation is further aggravated by the fact that changes in market demand contributes significantly to dynamically high inventory levels, which has a negative impact on product cost.

A third reason for the shift in emphasis to the supply chain is on account of the realization by most supply chain companies that by maximizing the performance of one department or function may lead to a less than optimal performance for the company as a whole. The purchasing function may negotiate a lower price on a particular product component, but the cost to produce the finished product, as a whole, may rise on account of consequent inefficiencies in the plant. Companies must look across the entire supply chain to gauge the impact of decisions in any one area.

Advanced Manufacturing Research, a Boston-based consulting firm, developed a supply chain model which emphasised material and information flow between manufacturers and their trading partners (Davis, 1995). They believed that greater awareness was required of the following areas

- Greater sharing of information between vendors and customers.
- Horizontal business processes replacing vertical departmental functions.
- Shift from mass production to customized products.
- Increased reliance on purchased materials and outside processing with a simultaneous reduction in the number of suppliers.
- Greater emphasis on organizational and process flexibility.
- Necessity to coordinate processes across many sites.
- Employee empowerment and the need for rules-based real time decision support systems.
- Competitive pressure to introduce new products more quickly.

Consequently, companies are streamlining all operations and minimising the time-to-customer for all their products. For these reasons, expertly managing the supply chain has become critical for most companies. As the vice president of product supply/customer service at Procter and Gamble put it, "Winning in the marketplace of the 1990s is going to require a far different kind of relationship - one that recognizes that the ultimate winners will be those who understand the interdependence of retailer/manufacturer business systems and who work together to exploit opportunities to deliver superior consumer value" (Drayer, 1994). Managers in companies across the supply chain take an interest in the success of the other companies. They work together to make the whole supply chain competitive. They have the facts about the market, they know a lot about competition, and they coordinate their activities with those of their trading partners. They use technology to gather information on market demands and exchange information between organizations. Critical to managing the supply chain is managing the link between each node within the chain to synchronize the entire supply chain.

1.6 Accelerated growth of interest in supply chains in the 21st century

It had become increasingly evident, especially since the beginning of the 21st century, that much international research effort has been expended on supply chain planning. In this regard the emphasis has been on enhancing the operational configuration and performance of supply chains, so that they may be aligned with both the planning requirements and the prevailing conditions of operational uncertainty. Some examples of previous supply chain, under uncertainty, research works, in this regard, include Grossmann and You (2008), Chen and Lee (2007), Awudu and Zhang (2012), Liu and Papageorgiou (2012), Guillen and Grossmann (2009), Al-Othman et al. (2008), Tsiakis and Papageorgiou (2007), Chen et al. (2007) and Chen and Lee (2004). Other examples of related supply chain under uncertainty research work include Optimisation of Production-Distribution Systems under Uncertainty, Li et al. (2008), Optimisation of Delivery Systems under Uncertainty, Bit al. (1992), and Optimisation of Decentralised Supply Chains, Raj and Lakshminarayanan, (2008).

The term, 'Uncertainty' was not applied consistently in all cases. Sometimes it referred to 'fuzzy uncertainty' or impreciseness, Carlsson and Fuller (2002), and represented by $<$, $\leq$, $>$, $\geq$ e.g. fuzzy market demand $<$, $\leq$, $>$, $\geq$ 1,500 t/yr and sometimes it referred to 'stochastic uncertainty', i.e. probability, e.g. market demand follows a Gaussian distribution with population age-group. In these two cases, expressions such as Fuzzy Optimisation, Stochastic Optimisation and Stochastic-Fuzzy Optimisation were coined to describe the procedures involved in determining the best operating solutions under those uncertain operating conditions. To add to the confusion, the planning term, 'Multi-Objectivity was occasionally added to these expressions whenever there was a planning requirement to consider multiple supply chain objectives, be they maxima, minima or a combination of both in nature, in solutions for uncertain supply chain operations, e.g. Multi-Objective-Fuzzy Optimisation, Multi-Objective-Stochastic Optimisation and Multi-Objective-Stochastic-Fuzzy Optimisation.

A supply chain optimisation capability that could cater for any prevailing conditions of operational uncertainty and that could also cater for any desired number of performance objectives would be of great interest to the commercial industrialised world where various combinational instances of operational uncertainty and corresponding planning requirements frequently occur. A good example of this is ammonia, (NH_3), from coal production and also the downstream production of fertiliser and explosives. Operational uncertainty is manifest in many ways. Firstly, with the production of NH_3 from coal, the concentration of hydrogen (H) in coal is probabilistically, or stochastically uncertain, which impacts not only upon ammonia production plant design considerations, but also upon the predictability of downstream product production, i.e. nitric acid (HNO_3) and ammonium nitrate (NH_4NO_3). Secondly, there is a need to maximise the production of NH_3 and downstream fertiliser and explosives products, in accordance with market demand, whilst simultaneously minimising the generation and discharge of hazardous production effluent, i.e. carbon monoxide (CO) and nitrogen dioxide (NO_2), which, taken together, are uncertain multi-objective planning requirements. (Note: this is an example of multi-objectivity where, one objective needs to be maximised while the other needs to be minimised. More frequently, all objectives need to be either maximised or minimised). Thirdly, the market demand for fertiliser product is imprecise, i.e. fuzzy ($<$, $\leq$, $>$, $\geq$) uncertainty. There are many other examples of commercial instances where multiple instances of planning and uncertainty exist, e.g. farming with downstream product beneficiation, the precious metal mining industry etc.

The availability of a planning and optimisation methodology for supply chains operating under any prevailing conditions of uncertainty would therefore be tantamount to being able to deliver ***'realistic'*** and planned supply chain operating solutions since all prevailing conditions of operational uncertainty would be accommodated.

In most business process or operational environments, there is usually a need to plan and function optimally, usually with regard to any number of operational objectives. However, since such operating environments are normally beset by problems of operational uncertainty, i.e. fuzzy uncertainty and stochastic (probabilistic) uncertainty, it is often very difficult to determine the best common operating conditions that will satisfy all process objectives, simultaneously and satisfactorily. Therefore, the opportunity is as follows:

An operational process, i.e. supply chain network, subject to prevailing conditions of operational uncertainty, i.e. fuzzy uncertainty and/or stochastic uncertainty, will undergo a realistically improved optimum performance should the impact of these uncertainty effects be taken into account and accommodated during the process design phase, irrespective of the planned number of operational objectives involved.

1.7 Planning with Uncertain Supply Chains

The primary purpose of this work is to define and derive a planning, i.e. multi-objective, and optimisation methodology for a supply chain network that is operating under conditions of uncertainty, i.e. fuzzy and stochastic (probability) uncertainty, and that may be successfully applied to any Supply Chain Network.

The secondary purpose of this initiative is to apply the derived methodology to a number of different supply chain infrastructural/operational configurations for the purpose of:

a) demonstrating how the methodology would be applied in different supply chain scenarios

b) demonstrating the different types of benefits that can be gained from the implementation of this new technology

Chapter 2

Concepts and Definitions

2.1 Supply Chain (Networks)

Prior to discussing the concepts of supply chain uncertainty and supply chain planning, it would be useful, at first, to agree on what exactly a supply chain (SC) or what a supply chain network is. Firstly, it is clear, according to the literature, that the terms, supply chain and supply chain network, both refer to precisely the same thing, and that their respective usage tends to be primarily dependent on whether the research, or other, focus, is based on supply chain (networks) in general or, specifically, on the networking aspects of supply chain infrastructures. A supply chain consists of a network of, *i)* production or manufacturing facilities with raw material and/or finished goods inventory holding capability (referred to as P in SCN diagrams), *ii)* distribution centres with warehousing (referred to as DC in SCN diagrams) and *iii)* Customer Zones (referred to as C in SCN diagrams), as is preferred in academic circles. Customer Zones are typically a collection of retail sales outlets in a specific geographic region with each customer zone experiencing a specific market demand for product. A typical supply chain network is depicted in the Fig. 2-1.

A review of the literature reveals that various definitions of supply chains or supply chain networks abound, with all having the same essential meaning but with slightly different nuances in application and also with varying degree of complexity. Some of the more common/popular definitions are described below:

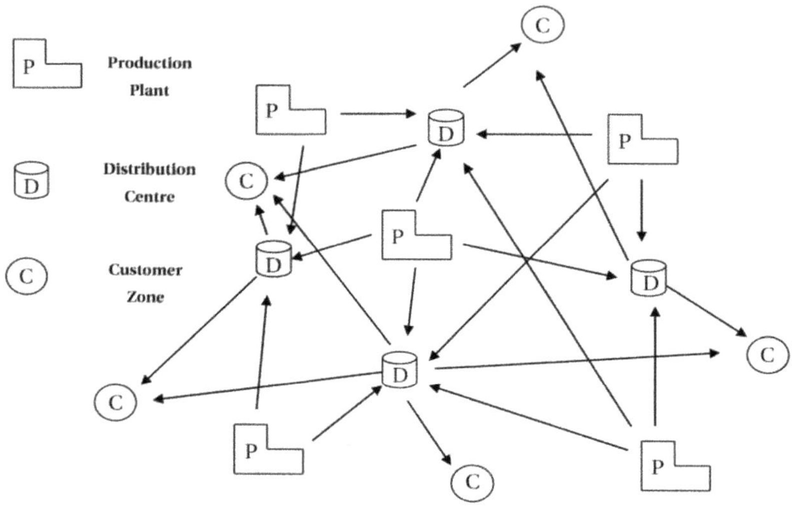

Figure 2-1Typical Supply Chain Network

i) Wikipedia

A supply chain is a system of organizations, people, activities, information, and resources involved in moving a product or service from supplier to customer. Supply chain activities transform natural resources, raw materials, and components into a finished product that is delivered to the end customer. Supply chains and supply networks both describe the flow and movement of materials & information, by linking organisations together to serve the end-customer.

ii) Investopedia

The network created amongst different companies producing, handling and/or distributing a specific product. Specifically, the supply chain encompasses the steps it takes to get a good or service from the supplier to the customer. Supply chain management is a crucial process for many companies, and many companies strive to have the most optimized supply chain because it usually translates to lower costs for the company. Quite often, many people confuse the term logistics with supply chain. In general, logistics refers to the distribution process within the company whereas the supply chain includes multiple companies such as suppliers, manufacturers, and the retailers.

iii) What is.com

A supply chain is the network of all the individuals, organizations, resources, activities and technology involved in the creation and sale of a product, from the delivery of source materials from the supplier to the manufacturer, through to its eventual delivery to the end user. The supply chain segment involved with getting the finished product from the manufacturer to the consumer is known as the distribution channel.

iv) Oxford Dictionary

The sequence of processes involved in the production and distribution of a commodity

v) Somewhere on the Internet

A supply chain is comprised of all the businesses and individual contributors involved in creating a product, from raw materials to finished merchandise.

vi) Somewhere else on the Internet

Supply chain management involves the planning of all activities as they relate to the flow of materials as they move from their source to the end customer, i.e. purchasing, manufacturing, labour management, warehousing, transportation and customer service.

From henceforth, for the sake of simplicity and comprehensiveness, the Oxford Dictionary definition of 'Supply Chain' will be applied, i.e. *'the sequence of processes involved in the production and distribution of a commodity'*. Now that a supply chain has been defined, it is possible to define supply chain uncertainty and supply chain planning.

2.2 Supply Chain, under Uncertainty and/or Planning, Optimisation Techniques

In order to properly conduct the required comprehensive literature research exercise on the existence of any previously researched, *singular*, e.g. multi-objective, fuzzy or stochastic optimisation, or *combinational*, supply chain, under uncertainty and/or planning, optimisation techniques, e.g. multi-objective-fuzzy or stochastic-fuzzy optimisation, these techniques must be clearly and properly defined up front. Two categories of supply chain, under uncertainty and planning, optimisation techniques are available and these are:

i) Singular Optimisation Techniques, e.g. fuzzy or stochastic optimisation

ii) Combinational Optimisation Techniques, e.g. multi-objective-fuzzy opt.

2.2.1 Singular Optimisation under Uncertainty or Planning Techniques currently available for Supply Chain Networks

It is clearly apparent, from most of the research literature listed in the Bibliography that the two instances of supply chain Uncertainty, i.e. fuzziness, stochastic uncertainty, and the one instance of supply chain planning, i.e. multi-objectivity, can exist in a supply chain operating environment. Further, it is also true that a few specific individual and combinational supply chain under uncertainty, optimisation techniques have been developed to provide optimum operating solutions for such environments, based on certain operating criteria, e.g. number and nature of suppliers, no. of distribution points etc. By *specific individual* optimisation methodologies is meant those techniques that apply to specific and definite areas of application in a supply chain operating environment, i.e. raw material procurement, production/manufacturing, distribution, sales. These various supply chain optimisation techniques include Fuzzy Optimisation, which means finding the best operating solution under uncertain operating conditions, and which is mathematically expressed as $<$, $\leq$, $>$, $\geq$. There is also Stochastic Optimisation, which means finding the best solution in a probabilistic environment, and which is usually expressed as $F(x_i)$).

Finally, there is Multi-objective Optimisation, which is the determination of the best operating solution in a planning environment where multiple operating objectives, either of a maxima, minima or combinational maxima and minima nature, can occur. Such specific individual uncertainty and/or planning optimisation methodologies are listed in Table 2-1.

The initial phase of this aspect of the research will be concerned with the identification and application of the various specific individual optimisation techniques such as Multi-Objective Optimisation Fuzzy Optimisation and Stochastic Optimisation, that can be found based on an appropriate literature search exercise. This will be followed by a study of the application of these various techniques with regard to the existence of any prevailing or limiting functional constraints in a typical process or production environment. The latter phase of this planning and optimisation study of supply chain under Uncertainty research will focus on the methods utilised in combining specific selected optimisation techniques into one desired methodology. Much recent research work has been conducted on combining any two specific planning and/or optimisation techniques into combinational techniques. These include Multi-Objective Fuzzy Optimisation, (Tsiakis and Papageorgiou (2007), Baky (2010)), Stochastic Fuzzy Optimisation, (Awudu and Zhang (2013), Chen and Lee (2004)), Multi-Objective Stochastic Fuzzy Optimisation (Grossmann and You (2008), Chen et al. (2007)), Bi-Objective Stochastic Fuzzy Optimisation (Gupta and Maranas (2003)) and Multi-Objective Stochastic Fuzzy (Pareto) Optimisation (Guillen an Grossmann (2010), Liu and Papageorgiou (2013)).

No evidence of research could be found into a supply chain planning and optimisation methodology that combines all three specific planning and uncertainty optimisation techniques into one *'Planning and Supply Chain Optimisation methodology under conditions of Uncertainty* methodology that would enable the determination of planned optimum supply chain solutions under any conditions of Uncertainty.

Table 2-1 Individual Supply Chain Optimisation Uncertainty/Planning Techniques Available

Optimisation Technique	Description
Fuzzy Optimisation	Finding the optimal decision in a set of crisp (real-valued) decisions and fuzzy membership function sets, $\{F_i \| i = 1 \text{ to } n\}$ or fuzzy optimisation selects from a set of real-valued, crisp, elements ordered by an aggregation operator on corresponding membership functions.
Multi-Objective Optimisation	Multi-objective programming, multi-criteria optimisation or *Pareto optimisation in special cases, is an area that is concerned with the mathematical optimisation of problems involving more than one objective function that have to be optimised simultaneously
Stochastic Optimisation	The process of maximising or minimising the value of a mathematical or statistical function when one or more of the input parameters is subject to randomness.

(*)Pareto Optimality: state of assignment of variables in which it is impossible to make any one function better off without making another worse off

The initial phase of this aspect of the research will be concerned with the identification and application of the various specific individual optimisation techniques such as Multi-Objective Optimisation Fuzzy Optimisation and Stochastic Optimisation, that can be found based on an appropriate literature search exercise. This will be followed by a study of the application of these various techniques with regard to the existence of any prevailing or limiting functional constraints in a typical process or production environment. The latter phase of this planning and optimisation study of supply chain under Uncertainty research will focus on the methods utilised in combining specific selected optimisation techniques into one desired methodology. Much recent research work has been conducted on combining any two specific planning and/or optimisation techniques into combinational techniques. These include Multi-Objective Fuzzy Optimisation, (Tsiakis and Papageorgiou (2007), Baky (2010)), Stochastic Fuzzy Optimisation, (Awudu and Zhang (2013), Chen and Lee (2004)), Multi-Objective Stochastic Fuzzy Optimisation (You and Grossmann (2008), Chen et al. (2007)), Bi-Objective Stochastic Fuzzy Optimisation (Gupta and

Maranas (2003)) and Multi-Objective Stochastic Fuzzy (Pareto) Optimisation (Guillen an Grossmann (2010), Liu and Papageorgiou (2013)).

No evidence of research could be found into a supply chain planning and optimisation methodology that combines all three specific planning and uncertainty optimisation techniques into one 'Planning and Optimisation of a Supply Chain Network under Uncertainty methodology that would enable the determination of planned optimum supply chain solutions under any conditions of Uncertainty.

2.2.1.1 Fuzzy Logic and Fuzzy Optimisation

According to the work of Carlsson and Fuller (2002), *Fuzzy logic* provides a logical framework that enables inexact human reasoning capabilities to be applied to knowledge-based systems. The theory of fuzzy logic provides a mathematical capability that enables the capture of those uncertainties normally associated with human learning processes, such as thinking and reasoning. Since traditional approaches to knowledge representation lack the means for representing the meaning of fuzzy concepts, and because subsequent approaches, based on first order logic, do not provide adequate representation of common sense knowledge, the development of fuzzy logic was motivated.

Further to the research work of Lodwick and Jamison (2007), fuzzy decision making or fuzzy optimisation involves finding the optimal decision in a set of crisp (real-valued) decisions and fuzzy membership function sets, $\{F_i|i = 1 \text{ to } n\}$ or fuzzy optimisation selects from a set of real-valued, crisp, elements ordered by an aggregation operator on corresponding membership functions.

2.2.1.2 Multi-Objective Optimisation

Further to the work of Duh and Brown (2007), multi-objective optimisation, also known as multi-objective programming, multi-criteria optimisation or Pareto optimisation in special cases, is an area that is concerned with the mathematical optimisation of problems involving more than one objective function that have to be optimised simultaneously. Multi-objective optimisation has been applied in many fields of science, including engineering, economics, operations management and logistics where optimal decisions need to be taken, typically in the presence of trade-offs between two or more conflicting objectives. Examples of multi-objective optimisation problems involving two and three objectives include, for example, maximising production of ammonia (NH_3) whilst simultaneously minimising the discharge of gaseous effluent, carbon monoxide (CO), maximising vehicle performance whilst simultaneously minimising fuel consumption and emission of pollutants and maximising the profitability, return on investment (ROI) and customer satisfaction levels of a company whilst simultaneously minimising working capital (e.g. cash at bank, debtors, inventory levels) expenditure.

Singular solutions may not exist that simultaneously and optimally satisfy all objective functions. Such objective functions are deemed to be conflicting in nature, and there will also be a possibility that an infinite number of Pareto optimal solutions may exist. A solution may be regarded as *non-dominated Pareto optimal, Pareto efficient* or *non-inferior* if none of the objective functions can be further improved in value without detrimentally influencing any of the other objective function values. Without any additional preference information, all Pareto optimal solutions may be regarded as equally valid, mathematically speaking. Researchers study multi-objective optimisation problems from different viewpoints, depending on their requirements, and therefore different solution philosophies and goals exist when defining and subsequently solving them. The goal may be finding a representative set of Pareto optimal solutions, and/or quantifying the trade-offs in satisfying the different objectives, and/or finding a single solution that satisfies the preferences of a human decision maker.

The context for multi-objective optimisation, in this discussion, is a supply chain operational environment.

2.2.1.3 Stochastic Optimisation

Further to the publication of Gibilisco and Isola (2011), stochastic optimisation is the process of maximizing or minimizing the value of a mathematical or statistical function when one or more of the input parameters are subject to randomness. The word stochastic means *involving chance or probability*.

Stochastic optimisation techniques involve those that apply random variables in generating optimum solutions. In the formulation of stochastic programs, random variables appear in the formulation of the optimisation problem itself, either in the form of objective functions or constraints. Stochastic optimisation methods can also involve problems with random variables generated by iteration, which may be used to solve certain stochastic problems. Stochastic optimisation techniques are based on deterministic methods for deterministic problems.

Again, the context for stochastic optimisation, in this discussion, is a supply chain operational environment.

2.2.2 Combinational Optimisation under Uncertainty Techniques currently available for Supply Chain Networks

As can be seen from Table 2.2, quite a few combinational uncertainty optimisation conditions are possible. For a comprehensive literature (journal) search, all possible combinations of all the various operational uncertainty terms available, i.e. multi-objective, fuzzy, stochastic, are considered in conjunction with a supply chain network infrastructure to produce the desired search results.

Table 2-2 Combinational Optimisation Methodologies

Combinational Optimisation Methodology	Application Conditions	Based on the work of:
Multi-Objective Fuzzy Optimisation	Finding the 'best' planning solution for a SCN under conditions of scalar uncertainty and having multiple operational objectives	Bit et al. (1992) Tsiakis & Papageorgiou (2007) Baky (2010)
Stochastic Fuzzy Optimisation	Finding the 'best' solution for a SCN Supply Chain Network) under conditions of stochastic fuzzy uncertainty	Gupta and Maranas (2003) Santoso et al. (2005) Awudu and Zhang (2013)
Multi-Objective Stochastic Fuzzy Optimisation	Finding the 'best' planning solution for a SCN under conditions of multiple objectivity and probabilistic uncertainty	Grossmann and You (2008) Guillen et al. (2005)
Multi-Objective (Pareto) Fuzzy Optimisation	Finding the 'best' planning solution for a SCN under conditions of fuzzy and multiple objective uncertainty, utilising the Pareto ε-constraint method	Guillen and Grossmann (2013) Liu and Papageorgiou (2013)

Research into possible, existing optimisation methodologies was conducted by performing literature research on the various combinational optimisation methodologies listed in Table 2-2. The results that were discovered are discussed for Stochastic Fuzzy Optimisation, Multi-Objective Fuzzy Optimisation and Multi-Objective Stochastic Fuzzy Optimisation from para. 2.2.2.1 to para. 2.2.2.3

2.2.2.1 Stochastic Fuzzy Optimisation

In the paper of Luhudjula and Gupta (1995), the applicability of optimisation models is extended to situations where both fuzzy and random data sets are involved. The major mathematical tool used to this end is the concept of fuzzy random variables, which provides a coordinated approach in dealing with these two kinds of imprecision. The extension principle is extended to fuzzy random variables and exploited to derive a way for handling robust fuzzy stochastic mathematical programming problems.

2.2.2.2 Multi-Objective Fuzzy Optimisation

The paper of Mahapatra and Roy (2002) considers multi-objective reliability optimisation problem for system reliability where reliability enhancement is involved with several mutually conflicting objectives. It is conflicting to reduce the cost of the system and improve the reliability of the same system simultaneously. A new fuzzy multi-objective optimisation method is introduced and it is used for the optimisation decision making of the series and complex system reliability with two objectives. In general, cost of the components of the system's objective goals has not been stated clearly. This imprecise reliability optimisation model is solved through the fuzzy multi-objective optimisation method. Alternatively, the research work of Bit et al. (1992), which deals with a 'Fuzzy programming approach to a multi-objective solid transportation problem', includes a well-defined fuzzy multi-objective optimisation methodology that can easily be applied, and, in fact, includes a numerical example.

2.2.2.3 Multi-Objective Stochastic Fuzzy Optimisation

In a paper presented by Bath et al. (2003) covering a multi-objective framework, an interactive fuzzy satisfying method, was presented to determine the optimum values of certain objectives subject to statistical uncertainties. In determining the optimum values, four objectives were simultaneously minimized. Specific techniques were put forward to convert the stochastic models into their respective deterministic equivalents. A weighting method was used to simulate the trade-off relationship between the conflicting objectives in the non-inferior domain.

The paper by Bath et al. (2004), a production generation schedule, involving NO/NO_2 pollutant emissions, system load demand and statistical uncertainties in production cost data is produced. In determining an optimum production schedule, four objectives viz. operating cost, NO/NO_2 emissions, variance of active risk and power generation mismatch are simultaneously minimised. Stochastic models are transformed into their respective deterministic equivalents by means of certain specific techniques. The trade-off between conflicting objectives is resolved by application of the weighting method. Contemporary search techniques, such as Hooke–Jeeves, are utilised to search the 'preferred' weightage pattern for a match, which would then correspond to an optimal solution. Fuzzy set theory is utilised in determining a 'preferred' optimal solution through the determination of maximum satisfaction levels of the participating membership functions. The nature of the results obtained demonstrates the validity of the proposed method.

2.3 Optimisation of Supply Chain Network under Uncertainty

2.3.1 Introduction

For all the literature uncovered on supply Chain Network Optimisation under Uncertainty, research was typically based on a multi-echelon supply chain network consisting of plants, distribution centres and customer zones as depicted in Fig 2-2.

Figure 2-2 Multi-Echelon Supply Chain Network

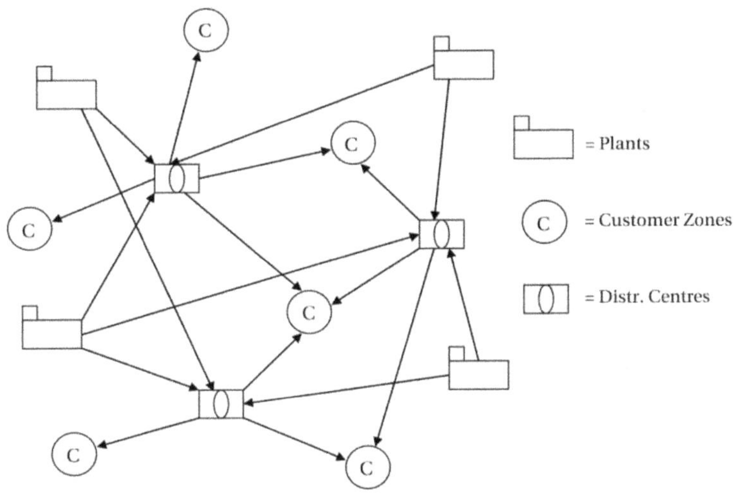

Similarly, based on an overview of <u>all</u> the journal papers listed under either 'Supply Chain Optimisation under Uncertainty' or 'Supply Chain Planning under

Similarly, based on an overview of <u>all</u> the journal papers listed under either 'Supply Chain Optimisation under Uncertainty' or 'Supply Chain Planning under Uncertainty' in the Bibliography, the fundamental keys stages involved in the Optimisation/Planning methodologies concerned are those listed in Table 2-4.

Table 2-4: Key aspects as to the optimisation of supply chain networks under Uncertainty

1	Illustration and description of the supply chain network (SCN) involved
2	Nature and degree of SCN operational Uncertainty involved
3	Description/Explanation of the optimisation methodology utilised.
4	Application by means of a case study.

Of these various stages involved, the one entitled, 'Nature and degree of SCN operational Uncertainty involved' requires further explanation, below.

2.3.2 Nature and degree of SCN operational planning and Uncertainty

As has been stated many times in this work, e.g. Liu and Papageorgiou (2013), Guillen and Grossmann (2009), Grossmann and You (2008), the operational performance of supply chain networks can be affected by the existence of conditions of operational Uncertainty, i.e. multi-objectivity, fuzziness and stochastic uncertainty in the supply chain network. Such conditions of operational uncertainty may occur either singly, e.g. fuzzy, stochastic uncertainty, or in a combinational fashion, e.g. multi-objective-fuzzy, multi-objective-stochastic-fuzzy uncertainty, in a supply chain network. The paragraphs below discuss some previously researched cases of both singular and combinational forms of supply chain operational uncertainty.

2.3.2.1 Supply chain network with a normal probabilistic demand distribution

This paper by Grossmann and You (2008) discusses the design and operational optimisation of Supply Chain Network under demand uncertainty, from a response time and economic perspective. The Supply Chain is composed of a multi-product, multi-site production and distribution network servicing multiple customer zones. The economic objective is measured in terms of net Present Value whilst the response time objective is measured in terms of transportation time, residence times, cyclic schedules in multi-product plants and inventory turnaround time, Expected lead time is proposed as the quantitative measure of Supply Chain responsiveness, which is achieved using a probabilistic model for stock-outs. Such probabilistic model can also be applied to predict the safety stock levels by integrating stock-out probability with demand uncertainty. These components are all built into a Multi-period mixed Integer Non Linear Programming (MINLP) model and the problem is formulated as a Bi-Criterion Optimisation model that maximises net present value and minimises expected lead time.

'Uncertainty in Market Demand' was researched by analysing three different research papers, which, although based on similar supply chain network infrastructures, they individually focussed on different approaches in handling Demand Uncertainty.

2.3.2.2 Supply chain network with discrete demand scenarios of known probabilities

In a research paper by Chen et al. (2007), the planning of a multi-product, multi-period and multi-echelon supply chain network that consisted of several existing plants at fixed places, a few warehouses and distribution centres at undetermined locations and a number of given customer zones. Unsure market demands were taken into account and modelled as a number of discrete scenarios with known probabilities. The supply chain planning model was constructed as a multi-objective mixed-integer linear program (MILP) to satisfy several conflicting objectives, such as minimising total cost, raising the decision robustness in various product demand scenarios, lifting the local incentives and reducing the transport time. For the purpose of creating a compensatory solution amongst all participants of the supply chain, a two-phase fuzzy decision-making method was presented and through the use of a practical, numerical example, an effective compromised solution was found for a multi-echelon supply chain network operating under uncertainty.

2.3.2.3 Supply chain network under Fuzzy Demand

In a research paper by Tsiakis and Papageorgiou (2007), it was stated that enterprise optimisation could rapidly strip significant 'bottom line' costs out of global operations, giving companies a real competitive edge. The benefits of managing supply chain networks by integrating operational, design and financial decisions have been acknowledged by the industrial and academic community.

The objective of this work was to determine the optimal configuration of a production and distribution network that was subject to operational and financial constraints. Operational constraints included quality, production and supply restrictions, and were related to the allocation of the production and the work-load balance. Financial constraints included production costs, transportation costs and duties for the material flowing within the network subject to exchange rates. As a business decision the out-sourcing of production was considered whenever the organisation could not satisfy the demand. A mixed integer linear programming (MILP) model was proposed to describe the optimisation problem. A case study for the coatings business unit of a global specialty chemicals manufacturer was used to demonstrate the applicability of the approach in a number of scenarios.

This paper had proposed an integrated model based on a detailed mathematical programming formulation that addressed some of the complex issues relating to the design and operation of global supply chain networks. The focus was on financial and tactical operational aspects within the organisation taking into account production balancing amongst sites. Between other business benefits was the operational and distribution efficiency of the network, visibility and control of the supply chain and capability to perform towards Key Performance Indicators (KPI's) such as operational cost, customer satisfaction and product quality. Moreover, the impact of decisions on the design and tactical operation could be quantified and evaluated. The data used was extracted directly from ERP systems, thus making such approaches easy to use and easy to update every time the application was used. The proposed MILP model was aimed at assisting senior operations management in making production allocation, production capacity per site, purchase of raw materials and network configuration decisions, taking into account financial aspects (exchange rates, duties, etc.) and costs.

2.3.2.4 *Supply chain network with a stochastic demand distribution plus case study*

In a research paper by Gupta and Maranas (2003), a two stage stochastic programming based approach was described to model the supply chain planning process as it reacts to demand realizations unfolding over time. In the proposed bi-level-framework, the manufacturing decisions are modelled as 'here-and-now' decisions, which are made before demand realization. Subsequently, the logistics decisions are postponed in a 'wait-and-see' mode to optimize in the face of uncertainty. In addition, the trade-off between customer satisfaction level and production costs is also captured in the model. The proposed model provided an effective tool for evaluating and actively managing the exposure of an enterprise's assets (such as inventory levels and profit margins) to market uncertainties. The key features of the proposed framework are highlighted through a supply chain planning case study.

Through a planning case study, the ability of the proposed framework to address key issues in managing uncertainties in supply chains was highlighted. It was shown that by utilizing the presented framework, a more realistic description of the total planning costs (in terms of a probability distribution in contrast to a point estimate) could be obtained. Consequently, this information could potentially be utilized to manage the risk exposure of the company's assets. Risk management initiatives aimed at reshaping this distribution such that the downside risk is minimized while maintaining the upside potential could be undertaken based on this information. To this end, the use of derivative financial instruments, such as options, futures and swaps, in conjunction with the developed framework is currently being investigated. In addition to controlling risk in the supply chain, the proposed framework was also shown to provide valuable insights into the customer relationship aspects of the supply chain.

2.3.2.5 *Supply chain network with stochastic demand distribution plus case study*

According to research work of Roghanian et al. (2007), Bi-level programming, a tool for modelling decentralized decisions, consists of the objective(s) of the leader at its first level and that of the follower at the second level. Three level programming results when second level is itself a bi-level programming. By extending this idea it is possible to define multi-level programs with any number of levels. In most of the real life problems in mathematical programming, the parameters are considered as random variables. The branch of mathematical programming which deals with the theory and methods for the solution of conditional extremum problems under incomplete information about the random parameters is called 'stochastic programming'. Enterprise-wide supply chain planning problems naturally exhibit a multi-level decision network structure, where for example, one level may correspond to a local plant control/scheduling/planning problem and another level to a corresponding plant-wide planning/network problem. Such a multi-level decision network structure can be mathematically represented by using 'multi-level programming' principles.

In this paper, a 'probabilistic bi-level linear multi-objective programming problem' is considered and its application in enterprise-wide supply chain planning problem where market demand, production capacity of each plant and resource available to all plants for each product are random variables and the constraints may consist of joint probability distributions or not. This probabilistic model is then converted into a deterministic equivalent at each level, to which a fuzzy programming technique is then applied to solve the resulting multi-objective nonlinear program and, in so doing, generating in an optimal compromise solution.

2.4 General Definitions

The definitions of certain terms that are infrequently used in this document are given below:

2.4.1 Monte Carlo simulation

A problem solving technique used to approximate the probability of certain outcomes by running multiple trial runs, called simulations, using random variables.

2.4.2 Lamont Boiler

High pressure forced water circulation boiler that has a large helical furnace wall and helical coils above and below.

2.4.3 SAA - Sample Average Approximation method

The methodology relies on approximating the underlying stochastic program via sampling, and solving the approximate problem via a specialized optimisation algorithm.

2.4.4 Benders decomposition algorithm

Benders Decomposition Algorithm: - a technique in mathematical programming that allows the solution of very large linear programming problems that have a special block structure. This structure often occurs in applications such as stochastic programming.

2.4.5 The Stackelberg leadership model

A strategic game in economics in which the leader firm moves first and then the follower firms move sequentially.

2.4.6 Lexicographic Min-Max (LMM) optimisation

This depends on searching for solutions minimal according to the lex-max order on a multidimensional outcome space. LMM is a refinement of the standard Min-Max optimisation in that apart from determining the largest outcome, the second largest outcome is also minimised.

Chapter 3

CONCEPTUAL FOUNDATION AND CORE RESEARCH CONCEPTS

3.1 Introduction

The conceptual foundation, or the key concepts, of this study can be derived from the Purpose statement (1.1), wherein it is stated: '*The primary purpose of this research is to define and derive a planning, i.e. multi-objective, and optimisation methodology for a supply chain network that is operating under conditions of uncertainty, i.e. fuzzy and stochastic (probability) uncertainty, and that can be successfully applied to any Supply Chain Network*'. Therefore, the core defining research terms/concepts in this statement are defined in Table 3-1

Table 3-1: Core research concepts

iii) Supply Chain Management, i.e. planning (multi-objectivity) and operations

iv) Supply Chain under Uncertainty, i.e. fuzzy and/or stochastic uncertainty, of which market demand, rather than supply, uncertainty is key

v) Supply Chain Program Modelling & Optimisation, i.e. MILP/MINLP programming and solution.

Following the categorisation and analysis of the selected research literature into the three research concepts listed in Table 3-1, the supply chain optimisation technique utilised within each will be extracted and documented in a common standardised format. Methodology extraction will be achieved in chapter 4 - '*Research and Development*' whilst in chapter 5 – '*Integration…*', the various independent supply chain optimisation techniques will be logically integrated, according to previously researched standards and norms, to create the desired, '*Planning and Optimisation methodology for a Supply Chain under Uncertainty*'

Note (1): The letter 'U' in the term 'Uncertainty' is capitalised because the term can be comprised of up to two uncertain components, i.e. Fuzzy uncertainty and/or stochastic uncertainty

3.2 Establishment of Literature Research portfolio based on degree of conformance with identified Core Research Concepts

Further to the definition of Core Research Concepts, Table 3-1, it is now possible to conduct a literature search exercise to identify potential journal research papers that conform to such Core Research Concepts. A comprehensive literature exercise was conducted utilising the Core Research Concepts, either singly or combinatorially, as search criteria. The large volume (~90) of research literature uncovered (Bibliography) was evaluated in terms of scope of operations and degree of fit to the selected research criteria, resulting in a final list of preferred research journal papers, Table 3-2, that are arranged according to research date, which is basically synonymous with the overall sequence, degree and extent of research that was conducted into the field of 'Supply Chain, under Uncertainty, Optimisation/Planning/ Management.

Table 3-2: Journal Research Literature conforming to Core Research Concepts

Literature Research Portfolio	Core Research Concepts		
	Supply Chain under (market demand) Uncertainty (Fuzzy, Discrete Probability, Continuous Probability, Stochastic Inventory Level based)	Supply Chain Network Management (Planning & Operations) under Uncertainty	SCN Program Modelling & Optimisation under Uncertainty (programming using multi-objective, fuzzy, stochastic objectives & constraints)
Gradient algorithm for chance constrained fuzzy non-linear goal programming (incl. transformation of probabilistic functions into deterministic equivalents) – based on the work of Lee and Olson (1985)		X	X
Multi-objective Fuzzy Optimisation of a Supply Chain under Uncertainty – based on the work of Bit et al. (1992)		X	X
Multi-objective Fuzzy Stochastic Optimisation of a SCN with probabilistic scenario's – Gupta and Maranas (2003)	X		X
Optimisation of a Supply chain Network under Stochastic Fuzzy uncertainty – based on the work of Santoso et al. (2004)	X		X

Optimisation of a Supply Chain Network under Multi-objective Stochastic Fuzzy Uncertainty – based on the work of Guillen et al. (2005)	X	X	X
Optimisation of a Supply Chain Network, under different configuration scenarios, and subject to conditions of Multi-object. Fuzzy Uncertainty – based on the work of Tsiakis and Papageorgiou (2007)		X	X
Optimisation of a Supply Chain Network under Multi-objective Stochastic Fuzzy Uncertainty based on the work of Chen et al. (2007)	X	X	X
Optimisation of a Supply Chain Network under Multi-objective Stochastic Fuzzy Uncertainty – based on the work of Grossmann and You (2008)	X	X	X
A Global (Pareto-) Optimisation Strategy for Chemical Supply Chains under uncertainty with environmental impact – based on the work of Guillen and Grossmann (2009)		X	X
The Solution of Multi-Objective Linear Programs through Fuzzy Goal Programming – based on the work of Baky (2010)		X	X
The Optimisation of a Bi-objective Supply Chain		X	X

under Fuzzy uncertainty – based on the work of Cadona-Valdés and Ozdemir (2011)			
The Optimisation of a Supply Chain Network (SCN) under Stochastic Fuzzy uncertainty – based on the work of Awudu and Zhang (2013)	X	X	X
Multi-Objective (Pareto-) Optimisation of Supply Chains in the process industry – based on the work of Liu and Papageorgiou (2013)		X	X
A bi-objective stochastic fuzzy model for a warehouse in a supply chain network – based on the work of Razmi et al (2013)		X	X

Note: *The absence of a marking (**X**) in a particular area does not necessarily indicate that no research was done in that particular area by those particular researchers, but rather that the emphasis of the research effort was in other marked areas, **X.** Further, research literature selection was also based on variation in uncertainty type as well as on variation in operational type.*

Such papers, constituting the final journal research list, will be analysed, and the supply chain, under uncertainty, optimisation techniques utilised therein will be extracted therefrom and then integrated to create the required and comprehensive 'Planning and Optimisation of a Supply Chain, under uncertainty, Optimisation' methodology. This portion of the work will be covered in the following chapter, 'Research and Development'.

3.3 Planning the Research

It was decided to conduct the literature research exercise in accordance with the *date* sequence of those selected research journal papers listed in Table 3-2. The basic reason for this is that such chronological date sequence best represents a logical and rational development path for the supply chain under uncertainty domain. Therefore this research would commence with a basic supply infrastructure consisting of only multiple origins and destinations, one product type, one instance of operational uncertainty, i.e. fuzzy uncertainty, and also a multi-objective planning requirement. Such research work was carried out by Bit et al. (1992) and was called, 'Fuzzy programming approach to multiobjective solid transportation problem'. On the other, later side of the spectrum would be the research papers by Awudu and Zhang (2013) and Liu and Papageorgiou (2013) which would discuss multiple instances of operational uncertainty, multi-objective Pareto optimality and with variation in supply chain infrastructure. Therefore the sequence of literature research to be conducted is reflected in Table 2-2.

Amongst the selected 'Supply Chain under Uncertainty' research works are a few that deal with the mathematical implications of certain supply chain scenarios, e.g. Lee and Olson (1985) that dealt with the transformation of probabilistic relationships into deterministic equivalents whilst the paper of Santoso et al. (2005) dealt with the provision of a higher number of probabilistic demand scenarios through the integration of a sampling strategy, SAA – Sample Average Approximation method, with an accelerated Benders decomposition algorithm. Despite the context of these papers being supply chain based, the primary focus, however, is on the mathematical treatment of the implications of certain supply chain scenarios. It is important that the proposed methodology be fully cognizant of the implications of these treatments so that they may be correctly handled and managed in the methodology. These papers will be dealt with last in the following R&D chapter.

Chapter 4

RESEARCH AND DEVELOPMENT

4.1 Introduction

It is the intention of this research to derive and then apply a 'Planning and Optimisation methodology for a Supply Chain under Uncertainty' for a ***generic*** process Supply Chain operating under any conditions of Uncertainty. By 'generic' is meant a supply chain infrastructure that is atypical of most supply chain applications in the commercial world and that can be easily extended to accommodate additional requirement, as depicted in Fig 4-1. Such supply chain may consist of multiple production sites (*P*), multiple distribution centres (*D*) and multiple customer zones (*C*) with many potential network interconnections between them.

In order to derive such methodology, it will be necessary to examine and analyse certain, selected and appropriate, previous supply chain under uncertainty research works that, individually or combinatory represent the fundamental aspects of generic supply chain architecture, as depicted in Fig. 4.1. It will also be necessary to examine and analyse those prior research works that represent the various individual or combinational aspects of operational planning and uncertainty that are evident in a supply chain operational environment (e.g. multi-objective, fuzzy, stochastic, multi-objective-fuzzy, stochastic-fuzzy, multi-objective-fuzzy-stochastic uncertainty etc.). This will be essential to fully understand how these mechanisms function and how they are applied in practise. It will also be very important that such analysis is cognisant of the integration and programming techniques used to compile any previously researched (full or partial) supply chain optimisation under (aspects of) uncertainty techniques, so that they may be successfully re-applied where necessary.

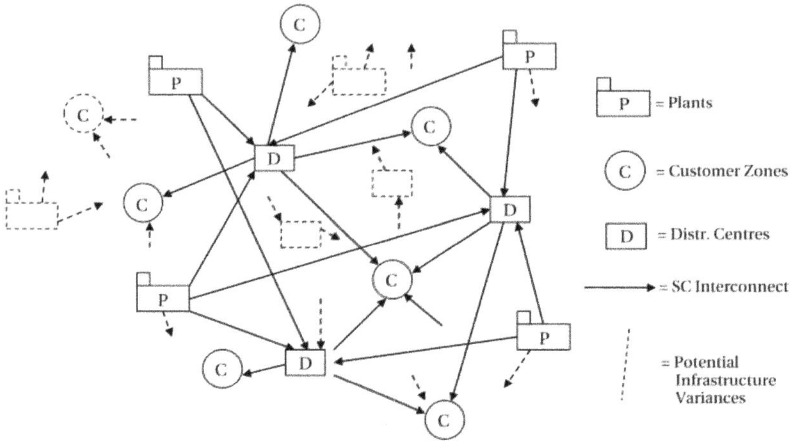

Figure 4-1 Generic Supply Chain Network that can easily be extended

Once the analysis of previously researched, optimisation of supply chain networks under individual or combinational conditions of uncertainty, is complete, it would then be possible to define and compile an Optimisation Methodology for a Supply Chain subject to conditions of operational uncertainty. Henceforth the research technique to be applied is itemised in Table 4-1, which is, in fact, a procedure that is common to most, 'Supply Chain, under Uncertainty, Optimisation/Planning/ Management literature researched, Table 3-2.

Table 4-1: Research Technique

a)	Diagrammatic description of generic Supply Chain Network
b)	Examination and analysis of previous and appropriate, Optimisation of Supply Chain under Uncertainty research works.
c)	Identification and extraction of all necessary elements for the creation of a Planning and Optimisation methodology for a Supply Chain Network operating under Uncertainty
d)	Definition of a Planning and Optimisation methodology for a Supply Chain Network operating under Uncertainty
e)	Integration and creation of a Planning and Optimisation methodology for a Supply Chain Network operating under Uncertainty
f)	Specification of a Planning and Optimisation methodology for a Supply Chain Network operating under Uncertainty
g)	Documentation

The Research and Development (R&D) work, described above and discussed below, will essentially follow the chronological, and *somewhat related*, development of the technology in the international R&D world. The expression 'somewhat related' is utilised because, although there are quite a few independent, but similar, research initiatives taking place concurrently around the world, at any point in time, there is frequent discussion and collaboration between the various research members and, consequently, various similarities will ensue.

The research literature identified in Table 2-2 will now be analysed and evaluated in accordance with the procedure outlined in Table 4-1.

4.2 Multi-objective Fuzzy Optimisation of a Supply Chain under Uncertainty

- based on the work of Bit et al. (1992) -

4.2.1 Introduction

A Fuzzy Multi-Objective Optimisation procedure for a supply chain under uncertainty was put together by Bit et al. (1992) and was termed, 'The Fuzzy Programming Approach to Multi-Objective Solid Transportation Problem'. Much prior research work had been done on the design of multi-objective transportation networks but Bit et al (1992) introduced a fuzzy programming algorithm for solving the multi-objective solid transportation problem. This is essentially a supply chain under multi-objective fuzzy uncertainty problem and it is the intention of this research to extract and formalise the multi-objective fuzzy optimisation technique so that it may be independently applied.

4.2.2 Problem Description

Further to the work of Bit et al. (1992):- A transportation network, consisting of m origins, O_i ($i = 1...m$) and n destinations, D_j ($j = 1...n$) is required to ship a_i units of a homogenous product to their n destinations in order to satisfy demand of b_j units of the product there. Let $c^p{}_{ijk}$ ($k = 1...K$) be the number of units of product which may be conveyed by K different modes of transport. A unit cost of $c^p{}_{ijk}$ is associated with the transportation of a unit of the product from source i to destination j by means of the k-th conveyance for the p-th criterion. However, in reality, the solid transportation problems are not of the singular objective type, they may involve more than one objective. Therefore a solid transportation problem may be represented as:

Minimise	$Z_p = \sum^{m}_{i=1}\sum^{n}_{j=1}\sum^{K}_{k=1} c^{p}_{ijjk}\, x_{ijk},$	$p = 1,\ldots,P$
subject to:	$\sum^{n}_{j=1}\sum^{K}_{k=1} x_{ijk}, = a_i$	$i = 1,\ldots,m$
	$\sum^{m}_{i=1}\sum^{K}_{k=1} x_{ijk}, = b_j$	$j = 1,\ldots,n$
	$\sum^{m}_{i=1}\sum^{n}_{j=1} x_{ijk}, = e_k$	$k = 1,\ldots,K$
	$x_{ijk} \geq 0,\ \ i = 1,., m, j = 1,\ldots,n,$	$k = 1\ldots K$

The subscript on Z_p and superscript on c^{p}_{ijk} denote the *P-th* cost criterion, $a_i > 0$ for all *i*, $b_j > 0$ for all j, $e_k > 0$ for all *k*, $c^{k}_{iijk} > 0$ for all *i, j, k, p*

$\sum^{m}_{i=1} a_i = \sum^{m}_{i=1} b_j = \sum^{K}_{k=1} e_k$ (balanced condition)

The balanced condition is both necessary and sufficient for the existence of a feasible solution to the transportation problem

4.2.3 Optimisation Technique for the Multi-objective Fuzzy uncertain Supply Chain environment

The multi-objective problem can be considered as a special case of the vector minimum problem whereby the first step is to assign, for each objective, Z_k, the following:

L_k = Aspiration Level of achievement for objective *k*.

U_k = Highest acceptable level of achievement for objective *k*.

Let $d_k = U_k - L_k$ = the degradation allowance for objective *k*

Once the aspiration levels and degradation allowance for each objective have been specified, the fuzzy model has been created; the next step is to transform the fuzzy model into a deterministic model.

i) Solve the linear programming problem by applying, sequentially, only one objective and ignoring the others

ii) From the results of step 1), determine the corresponding value for every objective for every solution derived, i.e. Zk(X), and Zk(Xk*) is the corresponding solution set.

iii) From step ii), determine, for each objective, the best (Uk) and the (Lk) values corresponding to the set of solutions, i.e.:

$U_k = \text{Max}\ \{Z_k(X^{1*}), Z_k(X^{2*}) \ldots, Z_k(X^{k*})\}$ and $L_k = Z_k(X^{k*})$

iv) A membership function, $\mu_k(Z_k)$ is then defined for each kth objective as follows:

$$\mu_k(Z_k) = \begin{cases} 1 & \text{if } Z_k \leq L_k \\ (U_k - Z_k)/(U_k - L_k) & \text{if } L_k < Z_k < U_k \\ 0 & \text{if } Z_k \geq U_k \end{cases}$$

v) A fuzzy linear program can then be formulated as follows::

Max λ_k, where $\lambda_k = (U_k - Z_k)/(U_k - L_k)$

subject to: $\Sigma^m_{i=1}\Sigma^n_{j=1}\Sigma^K_{k=1} cp_{ijk}x_{ijk} + \lambda_k)/(U_k - L_k)$, $k = 1, 2 \ldots K$

$\Sigma^n_{j=1}\Sigma^K_{k=1}\ x_{ijk}, = a_i$ $\quad i = 1,\ldots,m$

$\Sigma^m_{i=1}\Sigma^K_{k=1}\ x_{ijk}, = b_j$ $\quad j = 1,\ldots,n$

$\Sigma^m_{i=1}\Sigma^n_{j=1}\ x_{ijk}, = e_k$ $\quad k = 1,\ldots,K$

$x_{ijk} \geq 0,\ i = 1,., m, j = 1,\ldots,n,$ $\quad k = 1\ldots K$

,where $\lambda \geq 0$ and

vi) Solve the multi-objective fuzzy linear program under uncertainty using, for example, MS-Excel Solver, to determine the maximum value of λ for each objective function k

vii) The overall optimal compromise solution is given by the solution set that corresponds to the value of minimum value of max λ between them, i.e. $\lambda = \text{Min}_k(\mu_k(Z_k))$

4.2.4 Numerical Example

To illustrate the fuzzy programming algorithm discussed in Section 4.3.1.2, a multi-objective problem, having the following program characteristics is considered:

Minimise $Z_p = \Sigma^4{}_{i=1}\Sigma^4{}_{j=1}\Sigma^3{}_{k=1}\, c^p{}_{ijk}x_{ijk}$, $\quad p = 1, 2$

subject to

$\Sigma^4{}_{j=1}\Sigma^3{}_{k=1}x_{1jk} = 24$ $\quad \Sigma^4{}_{j=1}\Sigma^3{}_{k=1}x_{2jk} = 8$ $\quad \Sigma^4{}_{j=1}\Sigma^3{}_{k=1}x_{3jk} \leq 18$

$\Sigma^4{}_{i=1}\Sigma^3{}_{k=1}x_{i1k} = 10$ $\quad \Sigma^4{}_{i=1}\Sigma^3{}_{k=1}x_{i2k} \geq 11$ $\quad \Sigma^4{}_{i=1}\Sigma^3{}_{k=1}x_{i3k} \leq 19$

$\Sigma^4{}_{i=1}\Sigma^3{}_{j=1}x_{ij1} = 21$ $\quad \Sigma^4{}_{i=1}\Sigma^3{}_{j=1}x_{ij2} \geq 9$ $\quad \Sigma^4{}_{i=1}\Sigma^4{}_{j=1}x_{ij3} \leq 17$

$\Sigma^4{}_{i=1}\Sigma^4{}_{j=1}x_{ij2} = 31$ $\quad \Sigma^4{}_{i=1}\Sigma^4{}_{j=1}x_{ij3} = 12$

$x_{jk} \geq 0$, for $i = 1, 2, 3, 4$; $j = 1, 2, 3, 4$; $k = 1, 2, 3$

$C^1 = [c^1{}_{ijk}]$ and $C^2 = [c^2{}_{ijk}]$

$$C^1 = \begin{pmatrix} 15 & 18 & 17 & 12 & 22 & 13 & 10 & 4 & 12 & 8 & 11 & 13 \\ 17 & 20 & 19 & 21 & 21 & 22 & 21 & 19 & 18 & 30 & 10 & 23 \\ 14 & 11 & 12 & 25 & 34 & 33 & 20 & 16 & 15 & 21 & 23 & 22 \\ 22 & 18 & 13 & 24 & 35 & 32 & 18 & 21 & 14 & 13 & 23 & 20 \end{pmatrix}$$

$$C^2 = \begin{pmatrix} 6 & 7 & 8 & 10 & 6 & 5 & 11 & 3 & 7 & 10 & 9 & 6 \\ 13 & 8 & 11 & 12 & 2 & 9 & 20 & 15 & 13 & 17 & 15 & 13 \\ 5 & 6 & 7 & 11 & 9 & 7 & 10 & 5 & 2 & 15 & 14 & 18 \\ 13 & 6 & 6 & 17 & 11 & 18 & 12 & 16 & 12 & 18 & 14 & 7 \end{pmatrix}$$

Solution: Using MS Excel Solver

a) Program formulation

i) Objective Functions

$Z_1 = \text{Min}[15x_{111} + 18x_{112} + 17 x_{113} + 12x_{121} + 22x_{122} + 13x_{123} + 10x_{131} + 4x_{132} + 12x_{133} + 8x_{141} + 11x_{142} + 13x_{143} + 17x_{211} + 20x_{212} + 19 x_{213} + 21x_{221} + 21x_{222} + 22x_{223} + 21x_{231} + 19x_{232} + 18x_{233} + 30x_{241} + 10x_{242} + 23x_{243} + 14x_{311} + 11x_{312} + 12x_{313} + 25x_{321} + 34x_{322} + 33x_{323} + 20x_{331} + 16x_{332} + 15x_{333} + 21x_{341} + 23x_{342} + 22x_{343} + 22x_{411} + 18x_{412} + 13 x_{413} + 24x_{421} + 35x_{422} + 32x_{423} + 18x_{431} + 21x_{432} + 14x_{433} + 13x_{441} + 23x_{442} + 20x_{443}]$

$Z_2 = \text{Min}[6x_{111} + 7x_{112} + 8x_{113} + 10x_{121} + 6x_{122} + 5x_{123} + 11x_{131} + 3x_{132} + 7x_{133} + 10x_{141} + 9x_{142} + 6x_{143} + 13x_{211} + 8x_{212} + 11x_{213} + 12x_{221} + 2x_{222} + 9x_{223} + 20x_{231} + 15x_{232} + 13x_{233} + 17x_{241} + 15x_{242} + 13x_{243} + 5x_{311} + 6x_{312} + 7x_{313} + 11x_{321} + 9x_{322} + 7x_{323} + 10x_{331} + 5x_{332} + 2x_{333} + 15x_{341} + 14x_{342} + 16x_{343} + 13x_{411} + 6x_{412} + 6x_{413} + 17x_{421} + 11x_{422} + 18x_{423} + 12x_{431} + 16x_{432} + 12x_{433} + 18x_{441} + 14x_{442} + 7x_{443}]$

ii) Constraints

Equality Constraints	Equation		Limit
$^{4}{}_{j=1}{}^{3}{}_{k=1}X_{1jk} = 24$	24	=	24
$^{4}{}_{i=1}{}^{3}{}_{k=1}X_{i1k} = 10$	10	=	10
$^{4}{}_{i=1}{}^{3}{}_{j=1}X_{ij1} = 21$	21	=	21
$^{4}{}_{i=1}{}^{4}{}_{j=1}X_{ij2} = 31$	31	=	31
$^{4}{}_{j=1}{}^{3}{}_{k=1}X_{2jk} = 8$	8	=	8
$^{4}{}_{i=1}{}^{4}{}_{j=1}X_{ij3} = 12$	12	=	12

Inequality Constraints (≤)	-		
$^{4}{}_{j=1}{}^{3}{}_{k=1}X_{3jk} \leq 18$	1	≤	18
$^{4}{}_{i=1}{}^{3}{}_{k=1}X_{i3k} \leq 19$	19	≤	19
$^{4}{}_{i=1}{}^{4}{}_{j=1}X_{ij3} \leq 17$	12	≤	17

Inequality Constraints (≥)			
$^{4}{}_{i=1}{}^{3}{}_{k=1}X_{i2k} \geq 11$	11	≥	11
$^{4}{}_{i=1}{}^{3}{}_{j=1}X_{ij2} \geq 9$	31	≥	9

b) Optimisation of Objective Functions, individually, based on same set of constraints

i) Optimise function Z_1

Optimum Min $[Z_1] = 672$

Corresponding decision variables:

x_{111}	x_{112}	x_{113}	x_{121}	x_{122}	x_{123}	x_{131}	x_{132}	x_{133}	x_{141}	x_{142}	x_{143}
-	-	-	5	-	6	-	13	-	-	-	-

x_{211}	x_{212}	x_{213}	x_{221}	x_{222}	x_{223}	x_{231}	x_{232}	x_{233}	x_{241}	x_{242}	x_{243}
-	-	-	-	-	-	-	-	-	-	8	-

x_{311}	x_{312}	x_{313}	x_{321}	x_{322}	x_{323}	x_{331}	x_{332}	x_{333}	x_{341}	x_{342}	x_{343}
-	10	-	-	-	-	-	-	-	-	-	-

x_{411}	x_{412}	x_{413}	x_{421}	x_{422}	x_{423}	x_{431}	x_{432}	x_{433}	x_{441}	x_{442}	x_{443}
-	-	-	-	-	-	-	-	6	16	-	-

Corresponding value of $Z_2 = 659$ based on above set of decision variables

ii) Optimise function Z2

Optimum Min $[Z_2] = 359$

Corresponding decision variables

x_{111}	x_{112}	x_{113}	x_{121}	x_{122}	x_{123}	x_{131}	x_{132}	x_{133}	x_{141}	x_{142}	x_{143}
-	-	-	1	6	-	-	17	-	-	-	12

x_{211}	x_{212}	x_{213}	x_{221}	x_{222}	x_{223}	x_{231}	x_{232}	x_{233}	x_{241}	x_{242}	x_{243}
-	-	-	-	8	-	-	-	-	-	-	-

x_{311}	x_{312}	x_{313}	x_{321}	x_{322}	x_{323}	x_{331}	x_{332}	x_{333}	x_{341}	x_{342}	x_{343}
10	-	-	8	-	-	-	-	-	-	-	-

x_{411}	x_{412}	x_{413}	x_{421}	x_{422}	x_{423}	x_{431}	x_{432}	x_{433}	x_{441}	x_{442}	x_{443}
-	-	-	-	-	-	2	-	-	-	-	12

Corresponding value of Z_1 = 996 based on above set of decision variables

c) Formulation of Aspiration Level functions (to enable the determination of a single, non-conflicting multi-objective solution)

$$_k = \mu_k(Z_k) = (U_k - Z_k) / (U_k - L_k)$$

and $= \text{Min}\ (\mu_k(Z_k))$

Where U_k is the best level of achievement (highest value in a max situation or lowest value in a min scenario) and L_k is the worst level of achievement.

Therefore in this case:

Objective	U_k (Best)	L_k (Worst)
1	672	996
2	359	659

Consequently the following Aspiration Level functions may be derived:

$$_1 = 0.046x_{111} + 0.056x_{112} + 0.0525\ x_{113} + 0.037x_{121} + 0.068x_{122} + 0.04x_{123} + 0.031x_{131} + 0.012x_{132} + 0.037x_{133} + 0.025x_{141} + 0.034x_{142} + 0.04x_{143} + 0.052x_{211} + 0.062x_{212} + 0.058x_{213} + 0.065x_{221} + 0.065x_{222} + 0.068x_{223} + 0.065x_{231} + 0.059x_{232} + 0.056x_{233} + 0.093x_{241} + 0.031x_{242} + 0.071x_{243} + 0.043x_{311} + 0.034x_{312} + 0.037x_{313} + 0.077x_{321} + 0.104x_{322} + 0.102x_{323} + 0.062x_{331} + 0.05x_{332} + 0.046x_{333} + 0.065x_{341} + 0.071x_{342} + 0.068x_{343} + 0.068x_{411} + 0.056x_{412} + 0.04x_{413} + 0.074x_{421} + 0.108x_{422} + 0.099x_{423} + 0.056x_{431} + 0.065x_{432} + 0.043x_{433} + 0.04x_{441} + 0.071x_{442} + 0.062x_{443} - 2.074$$

$$_2 = 0.027x_{111} + 0.032x_{112} + 0.036x_{113} + 0.045x_{121} + 0.027x_{122} + 0.0225x_{123} + 0.05x_{131} + 0.0135x_{132} + 0.032x_{133} + 0.045x_{141} + 0.041x_{142} + 0.027x_{143} + 0.059x_{211} + 0.036x_{212} + 0.05\,x_{213} + 0.054x_{221} + 0.009x_{222} + 0.041x_{223} + 0.09x_{231} + 0.068x_{232} + 0.059x_{233} + 0.077x_{241} + .068x_{242} + 0.059x_{243} + 0.022x_{311} + 0.027x_{312} + 0.032x_{313} + 0.05x_{321} + 0.041x_{322} + 0.032x_{323} + 0.045x_{331} + 0.023x_{332} + 0.009x_{333} + 0.068x_{341} + 0.063x_{342} + 0.081x_{343} + 0.059x_{411} + 0.027x_{412} + 0.027x_{413} + 0.077x_{421} + 0.05x_{422} + 0.081x_{423} + 0.054x_{431} + 0.072x_{432} + 0.054x_{433} + 0.081x_{441} + 0.063x_{442} + 0.032x_{443} - 1.617$$

d) Maximisation of Aspiration Level functions (λ_k) based on the same set of constraints

Max $[\lambda_1] = 5.45$

Max $[\lambda_2] = 4.14$

The solution is given by $\lambda = \text{Min}\ [\lambda_k],\ \lambda_k \geq 0 = \text{Min}\ [5.45, 4.14] = 4.14$

The corresponding Objective Function values are Z_1 = 1,206 and Z_2 = 918 and the corresponding optimum decision variables are:

x_{111}	x_{112}	x_{113}	x_{121}	x_{122}	x_{123}	x_{131}	x_{132}	x_{133}	x_{141}	x_{142}	x_{143}
-	10	-	-	-	-	2	-	-	-	12	-
x_{211}	x_{212}	x_{213}	x_{221}	x_{222}	x_{223}	x_{231}	x_{232}	x_{233}	x_{241}	x_{242}	x_{243}
-	-	-	-	-	-	8	-	-	-	-	-
x_{311}	x_{312}	x_{313}	x_{321}	x_{322}	x_{323}	x_{331}	x_{332}	x_{333}	x_{341}	x_{342}	x_{343}
-	-	-	-	-	-	-	-	-	-	-	1
x_{411}	x_{412}	x_{413}	x_{421}	x_{422}	x_{423}	x_{431}	x_{432}	x_{433}	x_{441}	x_{442}	x_{443}
-	-	-	-	-	11	-	9	-	11	-	-

4.2.5 Extraction of the Supply Chain Optimisation Technique under Multi-objective Fuzzy Uncertainty

An analysis of the Paper, 'Optimisation of a Transportation Problem under Multi-objective Fuzzy Uncertainty' by Bit et al. (1992) reveals that the basic optimisation methodology for multi-objective fuzzy uncertainty employed may be summarised in Table 4-2 The relevant programming aspects of the paper are fully described in section 4.2.3

Table 4-2: Optimisation under Multi-objective Fuzzy Uncertainty Technique based on the work of Bit et al. (1992)

1 **Define the supply chain problem**

2 The requirement is to determine a compromise optimal solution amongst all operational objectives

3 Define Nomenclature requirements

4 Specify and define the objectives of the program

5 Specify and define the constraints and limitations for the program

6 Execute the Aspiration Level Multi-max or Multi-min subroutine

i) Solve the MINLP problem, applying, each time, only one objective and ignoring the others

ii) From the results of step i), determine the corresponding value for every objective for every solution derived

iii) From step ii), determine, for each objective (Zk), the best (Lk) and worst (Uk) values corresponding to the set of solutions. The 'best' (Lk) will be the highest value in a maximum scenario or lowest value (Uk) in a minimum scenario whereas the situation is reversed in a 'worst' case scenario

iv) Formulate Aspiration Level relationship whereby, $\lambda k < (Zk - Lk)/(Uk - Lk)$ for each objective function, k

v) Maximise λk in a MINLP for each objective function subject to constraints, typically using MS Excel Solver. The optimal compromise solution for the multi-objective problem is determined by the Aspiration Level (λk) function with the highest value

Maximise λk in a MINLP for each objective function subject to constraints

7 The optimal compromise solution for the multi-objective problem is determined by the Aspiration Level (λ_k) function with the highest value

8 Record optimal compromise solution

4.2.6 Conclusion

For the multi-objective fuzzy linear program with K objective functions, the fuzzy programing method described gives K non-dominating solutions that effectively provides for an optimal compromise solution. Furthermore, the optimisation technique utilised in this technique was extracted and is documented in Table 3-9.

4.3 Optimisation of a Supply Chain Network under Multi-objective Fuzzy Stochastic (probabilistic scenarios) Uncertainty

- based on the work of Gupta and Maranas (2003) -

4.3.1 Introduction

In accordance with the paper of Gupta and Maranas (2003), optimal supply chain network design can be achieved, under conditions of demand uncertainty, through the application of stochastic programming. With the proposed bi-level solution framework, the manufacturing decisions are modelled as '*here-and-now*' decisions that are made before demand realisation has occurred, whilst the logistics decisions are postponed in a '*wait-and-see*' mode until after demand realisation has happened. Cognisance is also taken of the trade-off between customer satisfaction levels and production costs in the model. In this way the exposure that exists, in terms of profit margins and inventory levels, is managed in accordance with market demand uncertainty. The key features of this proposed model are highlighted in a deterministic mid-term supply chain planning model, which effectively takes into account the overlap between these various phases.

4.3.2 Deterministic production model

A representative formulation of this supply chain under demand uncertainty case study will be achieved through the application of deterministic modelling. The intention behind this modelling approach is the determination of optimal sourcing and allocation of an enterprise's limited resources to its manufacturing assets so as to satisfy the market demands in the most cost-effective way. The supply chain network contemplated in the model consists of multiple production sites, manufacturing multiple products that are potentially located anywhere around the world. Market demand for these products exists at a variety of different customer locations situated around the world. The planning horizon for the model, which is in keeping with its midterm nature, ranges between one and two years. Each production site is represented by one or more single stage semi-continuous processing units, but with each having limited production capacity. Plant operations have to be managed to cater for the manufacture of various product sets that are part of various product groupings.

Overall operations can be apportioned into two distinct phases, i.e. the manufacturing phase and the logistics phase. The manufacturing phase is concerned with the efficient allocation of the production resources at the various production sites with the intention of establishing optimal production policies. Subsequently, in the logistics phase, various post-production activities such as the management of inventory levels as well as the management of customer satisfaction levels are undertaken in order to effectively meet customer demand. Consequently, the classification of supply chain activities, which, according to Gupta and Maranas (2003), may be translated into the (MP) model, described below. Nomenclature utilised is defined in Table 4-3.

MP

$$\text{Min } \sum_{fjst} FC_{fjs} Y_{fjst} + \sum_{ijst} v_{ijs} P_{ijst} + \sum_{ist} P_{is} C_{ist} + \sum_{iss't} t_{s's} W_{iss't} + \sum_{isct} t_{sc} S_{isct} + \sum_{ist} h_{ijs} I_{ist} + \sum_{ijst} \zeta_{js} I^{\triangle}{}_{ist} + \sum_{ijst} \mu_{ic} I_{ict} \quad (1)$$

subject to:

Manufacturing constraints:

$$P_{ijst} = R_{ijst} RL_{ijst} \quad (2)$$

$$C_{ist} = \sum_{i} \beta_{i'is} \sum_{j} P_{ijst}$$
(3)

$$C_{ist} = \sum_{s} W_{iss't} \quad (4)$$

$$FRL_{fjst} = \sum_{\lambda=1} RL_{ijst} \quad (5)$$

$$\sum_{f} FRL_{fjst} \leq H_{jst} \quad (6)$$

$$MRL_{fjs} Y_{fjst} \leq FRL_{fjst} \leq H_{jst} Y_{fjst} \quad (7)$$

$$A_{ist} = I_{is(t-1)} + \sum_{j} P_{ijst} - \sum_{s} W_{iss't} \quad (8)$$

Logistics constraints

$$I_{ist} = A_{ist} - \sum_{c} S_{isct} \quad (9)$$

$$\sum_{s,t \leq 1} S_{isct} \leq \sum_{t'<t} d_{ict} \quad (10)$$

$$I^{-}{}_{ic(t-1)} + d_{ict} - \sum_{c} S_{isct} \leq I^{-}{}_{ict} \leq \sum_{t'<t} d_{ict} \quad (11)$$

$$^{L}{}_{ist} - I_{ist} \leq I^{\triangle}{}_{ist} \leq I^{L}{}_{ist} \quad (12)$$

$$P_{ijst}, RL_{ijst}, FRL_{fjst}, C_{ist}, W_{iss't}, A_{ist}, I_{ist}, I^{-}{}_{ict}, S_{isct}, I^{\triangle}{}_{ist} \geq 0 \quad (13)$$

Table 4-3 Nomenclature for Multi-objective Fuzzy Stochastic modelling – Gupta and Maranas (2003)

Sets	
{i}	set of products
{f}	set of product families
{j}	set of processing units
{s}	set of production sites
{t}	set of time periods
Parameters	
FC_{fjs}	fixed production cost
v_{ijs}	variable production cost for product i on unit j at site s
p_{is}	price of raw material i at site s
$t_{ss'}$	transportation cost from site s to site s'
$t_{ss'}$	transportation cost from site s to site c
h_{ist}	Inventory holding cost for product i at site s in period t
ζ_{is}	safety stock violation penalty for product i at site s
μ_{ic}	revenue per unit of product i sold to customer c
R_{ijst}	rate of production of product i on unit j at site s in period t
$\beta_{i'is}$	yield adjusted amount of product i consumed to produce product i' at site s
λ_{if}	0 – 1 parameter indicating whether product i belongs to family f
H_{jst}	production capacity of unit j at site s in period t
MRL_{fjs}	minimum run length for family f on unit j at site s
d_{ict}	demand for product i at customer c in period t
I^{L}_{ist}	safety stock for product i at site s in period t
Variables	
Y_{fjst}	Binary variable indicating whether product family f is manufactured on unit j at site s in period t
P_{ijst}	production amount of product i on unit j at site s in period t
RL_{ijst}	run length of product i on unit j at site s in period t
FRL_{fjst}	run length of product f on unit j at site s in period t
C_{ist}	consumption of product i at site s in period t
$W_{iss't}$	Intersite shipment of product i from site s to site s' in period t
A_{ist}	amount of product i available for supply at site s in period t
I_{ist}	Inventory of product i from site s to customer c in period t
S_{isct}	supply of product i at site s to customer c in period t
I^{-}_{ict}	shortage of product i at customer c in period t
I^{Δ}_{ist}	deviation below safety stock of product i at site s in period t

With such formulation of MP, a deterministic viewpoint of supply chain management is taken with the understanding that all model parameters, such as cost coefficients, production rates, demand, etc. are known with complete certainty. However, this assumption of operational certainty regarding deterministic information, although desirable from a model complexity limitation perspective, is highly optimistic. However, operational uncertainty is more expected on account of the highly dynamic nature of most supply chains that are characterised by many sources of both technical and commercial uncertainty. Areas of concern such as rates of production and bill-of-material relationships, which are often subject to considerable variation, give rise to technical uncertainty in the manufacturing phase of the supply chain. Similarly, the exposure of the supply chain to commodity prices fluctuations and seasonal demand variations, gives rise to commercial uncertainty or market risk, which becomes evident in planning decisions. Within this range of various sources of uncertainty, probably the most important and extensively studied one is demand uncertainty. The emphasis on incorporating demand uncertainty into management considerations is understandable given the knowledge that the effective meeting of customer demand is primarily what drives most supply chain planning activities. Therefore, the development of a framework for incorporating demand uncertainty into the midterm planning model is considered important.

4.3.3 Incorporating uncertainty into the model

A most important consideration in accommodating uncertainties in supply chain planning decisions is establishing the appropriate representation of uncertain parameters. Two possible methods for representing uncertainty can be identified and these are the Scenario-based approach and the Distribution-based approach

i) Scenario-based approach

With the scenario approach, uncertainty is described by a set of discrete scenarios capturing how the uncertainty might play out in the future. Each scenario is associated with a probability level representing the decision maker's expectation of the occurrence of that particular scenario. Clearly however, the scenario-based approach is limited by the fact that it requires forecasting of all possible outcomes of an uncertain parameter

ii) Distribution-based approach

This technique involves the assignment of a probability distribution to a continuous range of potential outcomes

Once the possible techniques for representing uncertain parameters have been established, it then becomes possible to apply them appropriately in uncertain scenarios. In this regard, an enterprise can adopt two strategic 'positions' when faced with demand uncertainty. It can either position itself as a ***shaper*** or as an ***adapter*** to combat demand uncertainty. In the former case, the enterprise aims to restructure or 'shape' the demand distribution so that the associated downside risk is limited while the upside potential is retained. This is typically achieved through contracting agreements with the customer. With an *adapter* approach to the planning process, an enterprise does not attempt to influence the level of uncertainty in the market; it rather manages uncertainty by constantly 'adapting' to the risk exposure of its assets such as inventory levels and profit margins.

Following the establishment of this concept, a recognised framework for planning under uncertainty called, *two-stage stochastic programming* (2SMP), was developed. In this regard, the decisions and constraints of a system are classified into two different sets. In the first set, first-stage variables, also known as design or '*shaper*' variables are determined prior to the resolution of the underlying uncertainty. Dependent on these design '*here-and-now*' decisions and the realisations of the uncertain parameter, the second-stage or control variables are determined to optimise in the face of resolving uncertainty. These '*wait-and-see*' recourse decisions model how the decision maker '*adapts*' to the unfolding uncertain events. The presence of uncertainty is reflected by the fact that both the second-stage decisions as well as the second stage costs are probabilistic in nature. Once established, the objective then is to minimize the sum of the first stage costs, which are deterministic in nature, with the expected value of the second-stage costs, which are stochastic in nature. The classification of the decisions of the midterm planning model into manufacturing and logistics naturally fits into a two-stage stochastic programming framework, which is described below.

4.3.4 Production planning with demand uncertainty

The midterm production-planning model under demand uncertainty is, according to Gupta and Maranas (2003), formulated as the following two stage stochastic program (2SMP)

2SMP

$$\min \Sigma_{fjst} FC_{fjs}Y_{fjst} + \Sigma_{ijst}\, y_{ijs}P_{ijst} + \Sigma_{ijst}\, p_{is}C_{ist} + \Sigma_{iss't}\, t_{ss'}W_{iss't} + \ldots$$

P_{ijst}; C_{ist}; FRL_{fjst}; RL_{ijst};

$W_{iss't}$; $A_{ist}>0$; Y_{fjst} ε (0;1)

$$+\,E\left\{\begin{array}{l}\text{Min}\Sigma_{isct}\, t_{sc}S_{isct} + \Sigma_{ist}h_{ist}I_{ist} + \Sigma_{ist}\,\zeta_{is}\, I^{\Delta}{}_{ist} + \Sigma_{ict}\,\mu_{it}\, I^{-}{}_{ict}\\ \text{subject.to:}\\ \text{Eqns. (9) – (12)}\end{array}\right.$$

In formulating (2SMP), the manufacturing variables are considered as the first-stage, '*here-and-now*' or design decisions while the logistics decisions are modelled as the second stage '*wait-and-see*' or adaptor decisions. The objective function of the model (2SMP) is composed of two terms. The first term captures the costs incurred in the manufacturing phase. The second term quantifies the costs of the logistics phase, which is determined by applying the expectation operator (E) to an embedded optimisation problem. This recourse optimisation problem determines the optimal supply policies and inventory profiles given the production levels and demand realisations for the various products. The function of the expectation operator is to average over all possible demand realisations the costs incurred in the logistics activities. The constraints describing the supply chain are also partitioned within this framework. Eqns. (2) - (8) provide the restrictions for the production activities while Eqns. (9) - (12) describe the logistics planning phase. The coupling between these two phases arises through Eq. (9) that models the carryover of inventory given the amount available for supply from the production phase A_{ist} and the actual supplies to the various customers S_{isct}. Model (2SMP), thus, aims at determining the optimal production parameters in a supply chain environment by minimising the sum of manufacturing and expected logistics costs.

Through the use of Model (2SMP), an effective tool is provided for managing the risk exposure of an enterprise. The uncertainty in product demand is translated into the uncertainty in the logistics decisions through the second stage inventory management problem. This implies that inventory levels, supply policies, safety stock deficits and customer shortages are dependent on the first-stage, manufacturing decisions and such demand realised. A probability distribution can, therefore, be associated with each of these decisions. These distributions provide valuable information regarding the sensitivity of the firm's assets, such as inventory levels and profit margins, to the external demand uncertainty. For example, by utilising the predicted distributions for the inventory levels at the various sites in the supply chain, key strategic options such as capacity integration, expansion and disinvestments can be uncovered. Similarly, based on the variability in supply levels and the associated shortages at the customer sites, quotations of guaranteed service levels can be rendered to strengthen the competitive advantage of the enterprise in the marketplace.

4.3.5 Case study

To highlight the proposed framework for managing demand uncertainty in CPI planning, it is applied to a representative supply chain network shown in Fig. 4-2. The supply chain consists of six production sites manufacturing a total of 30 products. Products 1 through 10 are manufactured at site 1 and 2 while products 11 through 20 are manufactured at sites 3 and 4. The supply chain infrastructure is shown in Fig 4-2

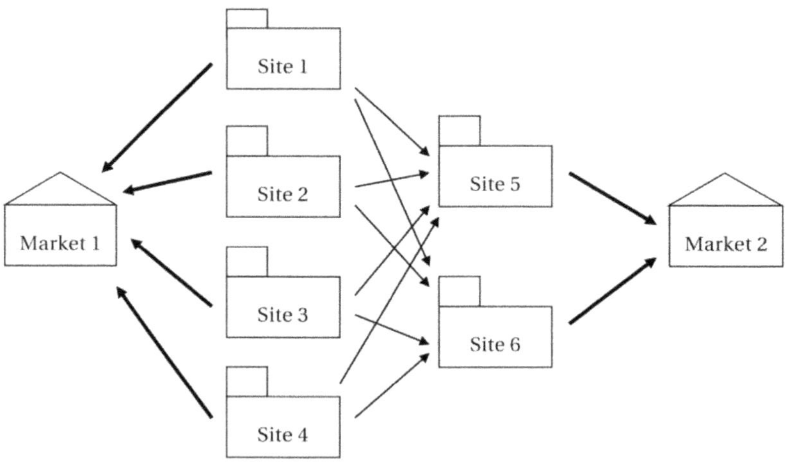

Figure 4-2: Supply chain configuration for the planning case study.

These 20 products are grouped into ten product families as follows: F1 = (1, 2); F2 = (3, 4); F3 = (5, 6). . . F10 = (19, 20). Thus, product families F1 - F5 are associated with sites 1/2 while F6 - F10 are associated with site 3/4. The demand for products 1- 20 exists externally at market 1 and internally at sites 5 and 6 where they are used as intermediate products. Specifically, products 21 - 25 are produced at site 5 while products 26 - 30 are manufactured at site 6. The demand for products 21 - 30 exists at market 2 as displayed in Fig. 4-2. An assembly-type product structure exists at sites 5 and 6, with one unit of each intermediate product being consumed to produce one unit of the final product. All the manufacturing sites are characterized by limited capacity processing equipment, which incur a set-up charge for each production campaign.

The demand is assumed to be normally distributed for each product with a coefficient of variation (standard deviation/mean) of 20%. The solution of model 2SMP for this supply chain setting results in an average total cost of 1512 units. The resulting convex MINLP is solved using a customized implementation of the outer approximation algorithm (OA) in approximately 2000 CPU seconds on an IBM RISC 6000 machine. The model statistics for the primal and master problem are: 1587 equations; 2351 variable s (primal); 2037 equations; 2861 variables; 480 discrete variables (master). In order to benchmark the computational efficiency achieved by solving the inner recourse problem analytically, the planning model is also solved using a Monte Carlo sampling method. A total of 500 scenarios are generated by sampling the underlying normal distribution and the resulting MILP (136 000 equations; 156 000 variables; 120 discrete variables) is solved using the CPLEX solver accessed via GAMS. The model is found to be extremely computationally intensive and fails to converge within the specified resource limit of 10 000 CPU seconds. This highlights the significant reduction in computational complexity that can be achieved by adopting the proposed solution technique over the naive, brute-force approach. Next, the impact of demand uncertainty on the various supply chain decisions was investigated. To this end, the optimal decisions obtained by solving model 2SMP were compared with those obtained by solving model MP. Table 4-11 lists how the capacity at sites 1 through 4 is allocated amongst the various product families.

Table 4-4: Capacity allocation to product families forecasted by 2SMP & MP models

Site	Product family (% capacity utilised)	Total Cap. Util. (%)
(a)	2SMP Model	
1	F2 (42%), F3 (32%), F4 (26%)	100%
2	F1 (50%), F5 (50%)	100%
3	F7 (14%), F9 (42%), F10 (30%)	86%
4	F6 (59%), F8 (36%)	95%
(b)	MP Model	
1	F1 (47%), F2 (37%)	84%
2	F3 (32%), F4 (19%), F5 (47%)	98%
3	F6 (54%), F10 (26%)	80%
4	F7 (15%), F8 (33%), F9 (29%)	77%

As expected, differences in allocation of capacity are observed both in terms of which product families are produced at a particular site as well as how much of capacity is allocated to them. The overall capacity utilization for each site is also predicted differently by the two models with the deterministic model consistently underestimating it as shown in Table 4-11. In the light of these differences, the question that is asked is how do the manufacturing decisions obtained by neglecting variability in product demand perform when exposed to uncertainty. To answer this question, a two-step procedure is applied. First, the deterministic model MP is solved. The resulting optimal first-stage manufacturing decisions are then fixed in model 2SMP and it is subsequently solved for the optimal second stage logistic decisions. This results in a total expected cost of 1557 units implying an increase of 2.9% over the original solution of model 2SMP. These results suggest that tangible cost savings can be realized through the incorporation of demand uncertainty in the planning process.

4.3.6 Extraction of the Optimisation of the supply Chain under Multi-objective Stochastic Uncertainty Technique

Table 4-5: Optimisation under Multi-objective Fuzzy Stochastic Uncertainty Technique based on the work of Gupta and Maranas (2003)

1	**Describe the supply chain network environment problem**
2	Describe the ideal operational requirements of the system - Optimum product portfolio and utilisation level per production plant - minimisation of total annualised costs of the network
3	Define the demand uncertainty:- Based on stochastic demand uncertainty, a two-stage stochastic program (2SMP) is formulated (should discrete, probabilistic scenarios have been evident, a scenario-based approach would have been followed)
4	Nomenclature Section
5	Formulate the 2SMP model by first partitioning the overall operations into 'here-and-now' (manufacturing) and 'wait-and-see' (logistics) phases
6	Formulate the first 'here-and-now' deterministic phase (manufacturing) MILP (mixed integer linear program) specify the objective function (minimisation of operating costs) define the constraints and limitations (2 – 8)
7	Formulate the second 'wait-and-see' uncertain phase MILP specify the objective function (minimisation of logistics costs) define the constraints and limitations (9 – 12) and apply the expectation operator, E[x] to average, over all demand realisations, the costs incurred in the logistics phase
8	Solve the first ' here-and-now' phase model by applying certain values, derived from the second 'wait-and-see' phase through Eqn. 9
9	Record the optimum results

4.3.7 Conclusion

In this paper, a deterministic model was adopted as the benchmark formulation for highlighting the various issues involved in incorporating uncertainty in the decision making process. Importantly, the supply chain network considered in this initiative was multi-product, multi-site and multi-period in nature. Other key aspects of the model included limited capacity production equipment and conflict between inventory and customer levels. It was also shown that a framework for incorporating demand uncertainty could be constructed by appropriately partitioning the decisions variables and constraints of the deterministic model. In particular, the supply chain operation was classified into manufacturing and logistics decisions. The manufacturing decisions were made before demand uncertainty was realised, while the logistics decisions were postponed until a later time. The delay of logistics decisions was used as recourse against the evolution of uncertainty in product demand. A case study was used to highlight the ability of the proposed framework to address the key issues of uncertainty management in supply chain planning activities. It was demonstrated that, a more realistic description of the total planning costs (in terms of a probability distribution in contrast to a point estimate) could be obtained by utilising such presented framework. The supply chain optimisation technique utilised herein was extracted and is recorded in Table 4-12.

Note: *A 2SMP supply chain modelling approach could be ideal for the production, distribution and sale of mass market commodity type items since management of product demand in those sectors tends to be based on inventory level dynamics.*

4.4 Optimisation of a Supply Chain Network under Multi-objective Stochastic Fuzzy Uncertainty

- based on the work of Guillen et al. (2005) -

4.4.1 Introduction

In a research paper by Guillen et al (2005), involving the integrated management of supply chains, the chief goals were the attainment of suitable economic returns for the Supply Chain (SC) stakeholders as well as the achievement of the desired satisfaction levels for customers. Additionally, since it was generally recognised that the operations of most supply chains are characterised by numerous sources of technical and commercial uncertainty, both deterministic and stochastic in nature, it was decided to adopt the ε-constraint method in conjunction with branch and bound techniques, to take into account not only SC profitability and customer satisfaction levels but also the uncertainty associated with financial risk, which is normally defined as the probability of not meeting targeted profit levels.

4.4.2 Problem Description

The chief intention of this research paper was, using the ε-constraint method in conjunction with branch and bound techniques, to determine the optimal design of a 3-echelon supply chain network (Fig 4-3) in terms of the maximisation of three objectives; NPV, Demand Satisfaction and Financial Risk. Optimal decisions to be taken include the capacity and location of plants and warehouses, the quantity of product to be manufactured at each plant and the flows of material

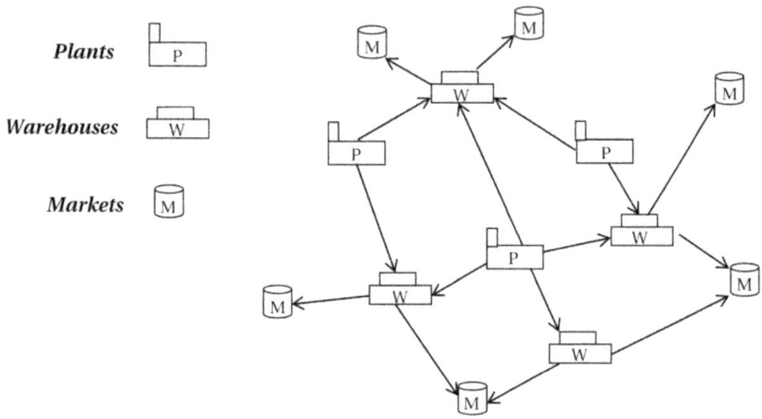

Figure 4-3: 3-echelon Supply Chain Network

4.4.3 Multi-objective Stochastic Model

A two stage stochastic programming approach is intended for this supply chain network in order to incorporate the uncertainty associated with demand within the modelling process. With such programming approach, the uncertain model parameters are considered as random variables with associated probability distributions and the decision variables are classified into two stages. The first stage variables are those that need to be realised prior to the incursion of uncertainty, whilst the second stage decisions can only be realised once the uncertainty is unveiled. On account of the stochastic nature of the second stage decisions, the objective function consists of the sum of the first stage performance measure and the expected value of second stage performance.

In this particular case, uncertainty associated with demand is represented by a set of scenarios, each with a given probability of occurrence. Such scenarios, in conjunction with their associated probabilities, constitute input data for the model. Moreover, those binary decision variables that characterise the network configuration through their representation of the different nodes of the supply chain network, and the continuous ones, which are related to the capacities of the sites, are jointly considered as first-stage variables because they need to be taken before the uncertainty is unveiled. On the other hand, however, decision variables related to the quantities of product to be produced and stored in the various nodes of the SCN are considered as second stage variables.

At the conclusion of each demand scenario, a different value of NPV and demand satisfaction is obtained for each particular realisation of demand uncertainty. The proposed model accounts for the maximisation of the expected value of the profit distribution, the target imposed for the customer satisfaction and the financial risk. The mathematical formulation of such model is next described. The supply chain modelling work, below, is based on the work of Guillen et al. (2005).

Table 4-6: Nomenclature for Multi-objective Stochastic Fuzzy Modelling – Guillen et al. (2005)

Subscripts	Physical Meaning
i	plants
j	warehouses
k	markets
o	objectives
p	products
s	scenarios
t	time intervals
Superscripts	**Physical Meaning**
L	lower bound of the variables
U	upper bound of the variables
Parameters	**Physical Meaning**
DE_{ts}	Direct expenses during t in s
Dem_{pkts}	Demand of p at k during t in s
Dep_t	Amount depreciated in t
$DRisk(x, \Omega)$	Downside risk associated with design x at target level Ω
$DSat_{ts}$	Total demand satisfaction during t in s
E[DSat]	Expected demand satisfaction
E[NPV]	expected net present value
FCI	fixed Capital Investment
FCI^L_{Pi}	fixed cost parameter at i
FCI^L_{WHj}	fixed cost parameter at j
HC_{pj}	unit handling cost of p at j
Ir	interest rate
IC_{pj}	unit inventory cost of p at j
IE^L_{pi}	indirect expense parameters at i
IE^L_{WHj}	indirect expense parameters at j
IE_t	indirect expenses during t
Inv_{pjts}	amount of p kept at j during t in s
$MDem_{pkt}$	mean demand of p at k during t
MDSat	demand satisfaction
nd	number of time intervals for depreciation
NPV	net present value
NPV_s	net present value in s
$PCap_i$	capacity of i
$PCap^L_i$	lower capacity bound of i
$PCap^U_i$	upper capacity bound of i
$Price_{pkt}$	price of p at k during t
$prob_s$	probability of s

Q_{pits}	amount of *p* produced at *i* during *t* in *s*
Rev_{ts}	revenues in *s* during *t*
$Risk(x, \Omega)$	risk associated with design x at target level Ω
$Sales_{pkts}$	sales of product *p* at *k* in *s* during *t*
SV	salvage value
tr	taxes
T	number of time intervals in the time horizon
Tax_{ts}	taxes in *s* during *t*
$TC1_{pij}$	unit transport cost of *p* between *i* and *j*
$TC2_{pik}$	unit transport cost of *p* between *j* and *k*
VC_{pi}	unit production cost of *p* at *i*
WC	working capital
$WHCap_j$	capacity of *j*
$WHCap^L_j$	lower capacity bound of *j*
$WHCap^U_j$	upper capacity bound of *j*
x	generic design variable
X_{pijts}	amount of *p* transported from *i* to *j* during *t* in *s*
Y_{pjkts}	amount of *p* transported from *j* to *k* during *t* in *s*
z_s	binary variable ($z_s = 1$ if $NPV_s < \Omega$, $z_s = 0$ otherwise)
Greek letters	Physical Meaning
γ_{pi}	fixed cost coefficient at *i*
γ_{WHj}	fixed cost coefficient at *j*
δ_s	auxiliary variable for downside risk definition
ε_0	bound level for 0
η_{pi}	indirect expense coefficient at *i*
η_{WHj}	indirect expense coefficient at *j*
λ_j	turnover inventory rate at *j*
μ	working capital factor
Ω	aspiration target level of profit

a) Mass balance constraints

The mass balance must be satisfied in each of the sites embedded in the SC. Therefore, for each plant i the total amount of product p manufactured during time interval t in scenario s (Q_{pits}) must be transported from i to warehouses

j as stated by Eq. (1), in which *Xpijts* represents the flow of p sent by i to j in t and s.

$$Q_{pits} = \sum_j X_{pijts} \text{f}\forall p, I, t, s \quad (1)$$

With regard to the warehouses, for each time interval t and scenario s, the total amount of product p sent by plant i to warehouse j (X_{pijts}) plus the initial stock of p kept at j at the beginning of t (Inv_{pjt-1s}) must be equal to the amount of p transported from j to final markets k (Y_{pjkts}) plus the final inventory of p in j (Inv_{pjts}), as expressed by constraint (2):

$$\sum_i X_{pijts} + \text{Inv}_{pjt-1s} = \sum_k Y_{pjkts} + \text{Inv}_{pjts} \quad \forall p, I, t, s \quad (2)$$

Moreover, it is considered that in the first period of time ($t = 1$), when the construction of the different sites of the SC is supposed to take place, the flows of materials between nodes are equal to zero as stated in constraint (3):

$$Q_{pits} = 0,\ X_{pijts} = 0,\ Y_{pjkts} = 0 \quad \forall p, i, j, k, t = 1, s. \quad (3)$$

Finally, the sales of product p carried out in market k during time interval t in scenario s (Y_{pjkts}) must be less than or equal to the demand (Dem_{pkts}) of this product at this market, (4):

$$\sum_k Y_{pjkts} \leq Dem_{pkts} \quad \forall p, k, t, s. \quad (4)$$

b) Capacity constraints

The capacity of each plant *i* is represented by a continuous variable (PC_{api}), which must be higher than the total amount of products *p* manufactured at *i* for every time interval *t* and scenario *s* as stated by Eq. (5). The parameter α_{pi} used in this expression weighs the amount of resources consumed by each product *p* at *i*.

$$\sum_p Q_{pits}\alpha_{pi} \leq \quad PC_{api} \quad \forall i, t, s. \tag{5}$$

Furthermore, PC_{api} is constrained by upper and lower bounds (PC^U_{api} and PC^L_{api}) *i*) in the event the plant is finally set as stated by Eq. (6). In this equation, *Ai* is a binary variable which takes the value of 1 in case the plant is opened and 0 otherwise. As can be observed, if the plant is not set, it cannot manufacture any product as its capacity is forced to assume a value equal to zero:

$$PC^L_{api} A_i \leq PC_{api} \leq PC^U_{api} A_i \quad \forall i \tag{6}$$

As for the plants, a continuous variable is defined to represent the capacity of the warehouses (WHC_{apj}). Therefore, the total inventory of *p* kept at warehouse *j* during time interval *t* in scenario *s* (Inv_{pjts}) must be lower than the capacity of the warehouse as stated by Eq. (7), where β_{pj} is a coefficient that weighs the storage resources consumed by each product *p* at *j*:

$$\sum_p Inv_{pjts}\, \beta_{pj} \leq WHC_{apj} \quad \forall i, t, s. \tag{7}$$

Moreover, the total flow of materials sent by warehouse *j* to the final market *k* is also constrained by the variable WHCap*j* as expressed by Eq. (8). In such expression, it is supposed that for each time interval *t*, scenario *s* and warehouse *j*, the capacity needed to handle a given amount of products, assuming regular shipment and delivery schedule, is twice the summation of the average inventory levels of products *p* (AIL_{pjts}) kept at *j* weighed by the storage coefficients, β_{pj}

$$2\sum_p AIL_{pjts}\, \beta_{pj} \leq WHCap j \quad \forall i, t, s. \tag{8}$$

AIL_{pjts} is computed by means of Eq. (9), in which λ_j represents the turnover ratio of the warehouse *j*, i.e., the number of times that the stock is completely replaced per time interval:

$$AIL_{pjts} = \sum_k Y_{pjkts} / \lambda_j \quad \forall p, j, t, s. \tag{9}$$

Finally, the capacity of the warehouse *j* is constrained by lower and upper bounds ($WHCap^L_j$ and $WHCap^U_j$) in case the warehouse is finally set, as stated by Eq. (10). The binary variable B_j used in this expression represents the existence of a warehouse *j* and takes the value of 1 in case *j* is opened and 0 otherwise.

$$WHCap^L_j\, B_j \leq WHCap_j \leq WHCap^U_j\, B_j \tag{10}$$

c) Objective function

The production/distribution system whose model has been described before must attain three targets:

a) maximise the NPV;

b) maximise the demand satisfaction, which in turn may bring future sales;

c) minimise the financial risk.

i) Net present value

As has been mentioned previously, different NPV values are obtained for each scenario under study (NPV_s) once the uncertainty is unveiled. The model described before must account for the maximisation of the expected value ($E[NPV]$) of the resulting NPV distribution, which can be computed by performing an average of the aforementioned NPV*s* as stated in Eq. (11):

$$E[NPV] = \sum_s prob_s\, NPV_s \tag{11}$$

The values of NPV*s* must be determined according to applicable rules (depreciation), legislation (taxes), etc., so this may lead to different formulations. The following equations intend to reflect a general case. Having in mind such purpose, NPV_s is calculated for each scenario s as the summation of the discounted cash flows (CF_{ts}) generated in each of the time intervals t in which the time horizon is divided as expressed by Eq. (12):

$$NPV_s = \Sigma_t CF_{ts}/(1 + ir)^{t-1} \tag{12}$$

With regard to the cash flows, these are computed for each scenario s and time interval t as the difference between the revenues (Rev_{ts}) and the total costs. Such costs include the direct (DE_{ts}) and indirect expenses (IE_t) as well as the taxes (Tax_{ts}) originated by all the production, distribution and storage activities performed in the day to day SC operation. Moreover, it is supposed that the necessary capital investment (CI) for carrying out the construction of the SC takes place in the first period of time, while the working capital (WC), which is part of this initial investment, and the salvage value of the SC (SV) are recovered in the last time interval as indicated by Eqs. (13) – (15):

$$CF_{ts} = -CI = -(FCI + WC)\ \forall s,\ t = 1 \tag{13}$$

$$CF_{ts} = -Rev_{ts} - DE_{ts} - IE_t - Tax_{ts} \quad \forall s,\ 1 < t < T \tag{14}$$

$$CF_{ts} = -Rev_{ts} - DE_{ts} - IE_t - Tax_{ts} + WC + SV \quad \forall s,\ t = T \tag{15}$$

The total capital investment (CI) is calculated by adding the fixed capital investment (FCI) and the working capital (WC). The first term includes the cost of setting the plants and warehouses, embedded in the supply chain network, which is a linear function of their capacities (16).

$$FCI = \Sigma_i (FCI^L{}_{Pi} A_i + PCap_i\, \lambda_{Pi}) + \Sigma_j (FCI^L{}_{WHj} Bj + WHCap_j\, \gamma_{WHj}) \tag{16}$$

On the other hand, the working capital (WC), which represents the initial amount of money that is necessary in order to initiate the production and distribution activities in the network, should be directly proportional to FCI as expressed by constraint (17). In such equation, the proportionality constant μ is considered to be equal to 0.194

$$WC = \mu FCI \tag{17}$$

The revenues obtained in each time interval t and scenario s are proportional to the sales of products p at markets k (Sales_{pkts}) and their associated prices (Price_{pkt}) as defined by Eqs. (18) and (19):

$$\text{Rev}_{ts} = \sum_{pk} \text{Sales}_{pkts} \, \text{Price}_{pkt} \quad \forall t, s, \tag{18}$$

$$\text{Sales}_{pkts} = \sum_{j} Y_{pjkts} \quad \forall p, k, t, s \tag{19}$$

With regard to the indirect expenses generated in each time interval t (IE_t), it is considered that these are proportional to the capacities of the plants/warehouses, as expressed by constraint (20).

$$\text{IE}_t = \sum_I (\text{IE}^L{}_{Pi} A_i + \text{PCap}_i \eta_{Pi}) + \sum_j (\text{IE}^L{}_{WHj} B_j + \text{WHCap}_j \, \eta_{WHj}) \; \forall t. \tag{20}$$

It should be mentioned that this term is computed directly from the first-stage variables (SC configuration) and without evaluating second-stage decisions, and therefore remains constant for all the scenarios under study once the uncertainty is unveiled. The direct expenses obtained in time interval t and scenario s ($\text{DE}ts$) are proportional to the amount of products manufactured, stored and transported through the different nodes of the SC, as stated by Eq. (21). Such costs include, therefore, the following terms:

- Variable production costs at the plants, which are assumed to be equal to the production rates of products p manufactured at plants i in each time interval t and scenario s (Q_{pits}) multiplied by the unit production costs (VC_{pi}).
- Handling costs at the warehouses, which in this case are assumed to be equal to the flows of products p sent by warehouses j to markets k in time interval t and scenario s (Y_{pjkts}) multiplied by the unit handling costs (HC_{pj}).
- Transport costs, which are supposed to be equal to the flows of materials transported between plants and warehouses (X_{pijts}) and warehouses and markets (Y_{pjkts}) in each scenario s and time interval t multiplied by the unit transport costs (TC1_{pij} and TC2_{pjk} respectively).

- Inventory-holding costs at the warehouses, which are supposed to be equal to the average inventory levels of product *p* kept at warehouse *j* in each scenario *s* and time interval *t* (AIL_{pjts}) multiplied by the unit inventory costs (IC_{pj}).

$$DE_{ts} = \Sigma_{pi}\, Q_{pits} VC_{pi} + \Sigma_{pjk}\, Y_{pjkts} HC_{pj} + \Sigma_{pij} X_{pijts} TC1_{pij} +$$

$$\Sigma_{pjk} Y_{pjkts} TC2_{pjk} + \Sigma_{pjk}\, Y_{pjkts} IC_{pj} / \lambda_I \quad \forall t, s \qquad (21)$$

Finally, the taxes to be paid in each time interval *t* and scenario *s* (Tax_{ts}) are computed assuming a linear depreciation policy as stated by constraints (22) – (24). In these equations tr represents the taxes rate to be applied on the gross benefit, and nd the number of time periods through which the depreciation will be carried out:

$$Tax_{ts} = (Rev_{ts} - Dep_t)\, tr \quad \forall 1 < t \leq nd + 1, s, \qquad (22)$$

$$Tax_{ts} = Rev_t\, tr \quad \forall nd + 1 < t \leq T, s \qquad (23)$$

$$Dep\, t = (FCI - SV) / nd \quad \forall t. \qquad (24)$$

ii) Demand satisfaction

Customer demand satisfaction for each time interval *t* and scenario *s* ($DSat_{ts}$) is measured as the average of the not covered demand (25):

$$DSat_{ts} = \Sigma_{pk}\, Sales_{pkts} / \Sigma_{pk}\, Dem_{pkts} \quad \forall t > 1, s. \qquad (25)$$

As with NPV, it is possible to compute the expected value of the demand satisfaction by performing an average of the values of $DSat_{ts}$ for all the scenarios *s* and time intervals *t* as expressed by constraint (26):

$$E[DSat] = \Sigma^{T}_{t=2} \Sigma_s\, prob_s\, Dsat_{ts} / (T - 1) \qquad (26)$$

The SC design problem, without considering the financial risk, would be therefore mathematically formulated as follows:

maximise {*E*[NPV]; *E*[DSat]}

subject to Eqs: (1) - (26).

The main drawback of such mathematical formulation lies on the fact that it does not reflect a realistic operational policy in terms of the demand satisfaction level to be achieved by the SC. If the expected value of the demand satisfaction is pursued as objective, the resulting Pareto optimal SC configurations may exhibit in some scenarios and time intervals demand satisfaction levels under the average while in others may exceed it. This means that the operational strategy of the SC towards the demand satisfaction will depend on the scenario that finally materialises, as well as the time interval considered, which does not seem desirable from the decision-maker's perspective. In order to overcome such difficulty and explicitly take into account the demand satisfaction strategy of the enterprise, a *minimum target* for the demand satisfaction (MDSat), which must be attained in all the time intervals and scenarios, is incorporated as an objective within the existing formulation, thus avoiding the use of the *E*[DSat]. This new issue leads to the following model:

maximise {*E*[NPV]; MDSat}

subject to Eqs. (1) - (25), where:

$$\text{MDSat} \leq \Sigma_{pk}\,\text{Sales}_{pkts}/\Sigma_{pk}\,\text{Dem}_{pkts} \quad \forall t > 1, s \qquad (27)$$

Therefore, by selecting a certain value for MDSat, it is guaranteed that for each scenario s and time interval $t > 1$ at least the minimum desired demand satisfaction level is reached. The formulation described above is thus able to reflect the operational policy that consists in obtaining the actual maximum profit at each scenario and time interval while ensuring that the demand satisfaction previously fixed is also achieved in all of them. Each SC configuration leads to two histograms, one for the NPV and another one for the demand satisfaction, and therefore exhibits an *E*[NPV] for a given operational policy (MDSat). This value of *E*[NPV] may change depending on the demand satisfaction target imposed to the SC operation, as will be discussed in the case study.

iii) Financial risk

The financial risk associated with a design project under uncertainty is defined as the probability of not meeting a certain target profit (maximisation) or cost (minimisation) level. For the two-stage stochastic problem, the financial risk associated with a design x and target profit Ω is therefore expressed by the following probability:

$$\text{Risk}(x, \Omega) = P(\text{NPV}(x) < \Omega) \tag{28}$$

, where $\text{NPV}(x)$ is the NPV after the uncertainty has been unveiled and a scenario realised. The definition of $\text{Risk}(x, \Omega)$ can be rewritten with the help of binary variables as follows:

$$\text{Risk}(x, \Omega) = \textstyle\sum_s \text{prob}_s\, z_s(x, \Omega) \tag{29}$$

, where z_s is a new binary variable defined for each scenario as follows:

$$z_s(x, \Omega) = \begin{cases} 1 & \text{if NPV}_s < 0 \\ 0 & \text{otherwise} \end{cases} \tag{30}$$

In a discrete scenario case, financial risk is given by the cumulative frequency obtained from the NPV histogram as depicted in Fig. 2. A more straightforward way of assessing and understanding the trade-offs between risk and profit is to use the cumulative risk curve as depicted in Fig. 3. A possible way of avoiding the use of binary variables to determine the risk consists of reformulating the problem without explicitly using this definition. For this purpose, the use of the concept of downside risk is applied. The $\text{DRisk}(x, \Omega)$ is calculated with the help of the following constraints:

$$\begin{aligned} &\text{DRisk}(x, \Omega) = \textstyle\sum_s \text{prob}_s\, \delta_s(x, \Omega), \\ &\qquad \delta_s(x, \Omega) \geq \Omega - \text{NPV}_s \quad \forall s, \\ &\qquad \delta_s(x, \Omega) \geq 0 \quad \forall s \end{aligned} \tag{31}$$

,where δ_s is a continuous variable. DRisk(x, Ω) can be utilised to control financial risk at different NPV targets by varying Ω from small values up to higher values and obtaining a full spectrum of solutions to be used by the decision-maker as a decision support tool.

The resultant objective function which includes the three objectives (NPV, demand satisfaction and financial risk) can be finally expressed as follows:

$$\text{maximise } \{E[\text{NPV}];\ \text{MDSat};\ -\text{DRisk}\} \tag{32}$$

The solution of this problem consists of a set of Pareto optimal SC configurations. These are obtained in this work by applying the ε-constraint method, first introduced by Haimes et al. (1971). This method is based on the maximisation of one objective function, and considering the other objectives as constraints bounded by some allowable ε-values. Such ε-values may be altered to generate the entire Pareto optimal set. Consequently, the following single MILP optimisation formulation is applied to obtain the Pareto solutions:

Maximise $E[\text{NPV}]$

subject to:

Eqs (1) - (24), (27) and (31)

$\text{MDSat} \geq \varepsilon_1$

$\text{DRisk}(x, \Omega) \leq \varepsilon_2$

Therefore, by changing the values of the bound levels ε_1 and ε_2, as well as the target Ω, a set of results can be obtained. Each of these results implies an SC configuration. The resulting Pareto solutions might be represented in a three dimensional chart (E[NPV], MDSat and DRisk(x, Ω)). The procedure to solve the proposed problem is as follows:

(1) Select a target for the downside risk calculation (Ω).

(2) Set initial targets for each objective (ε_1 and ε_2).

(3) Solve the proposed model.

(4) Obtain the corresponding Pareto solution.

(5) Choose a configuration from the Pareto set.

If the decision-maker is satisfied with the design then the process is terminated, otherwise step 1 follows. The proposed strategy should lead to a final SC design which would represent the desired compromise among the different objectives from the decision-maker's perspective. Regarding financial risk, it is also important to point out that this term can be managed by changing both, the target associated to the downside risk itself (ε_2) and the aspiration level for which such term is computed (Ω).

4.4.4 Extraction of the Optimisation under Multi-objective Stochastic Fuzzy Uncertainty Technique

Table 4-7: Optimisation of a supply chain under a Multi-objective Stochastic Fuzzy Uncertainty Technique based on the work of Guillen et al. (2005)

1	Problem description of the supply chain network including supply, production, warehousing, distribution and sales
2	Illustrated diagram of supply chain network
3	Requirements Definition: Determine the optimal configuration of the supply chain network in terms of the supply chain infrastructure and production rates of each product per plant
4	Define type of Uncertainty involved and Solution approach: Demand Uncertainty is represented by a set of scenarios with given probability of occurrence
5	Specify the constraints Mass balance constraints Capacity constraints
6	Specification of the Objective Function Maximise NPV Maximise the demand satisfaction Minimise the financial risk
7	Specification of Multi-objective Stochastic problem Maximise[E[NPV]; MDSat; -DRisk] per scenario Subject to constraints, eqns. (1) – (24), (27) and (31) MDSat ≥ 0 DRisk$(x, \Omega) \leq \varepsilon_2$
8	Compute the average NPV over all scenario's, $E[NPV] = \sum_s prob_s NPV_s$ Compute the expected value of demand satisfaction over all scenario's, $E[DSat] = \sum_{t=2}^{T} \sum_s prob_s Dsat_{ts} / (T-1)$ Determine the risk over all scenario's, $DRisk = \sum_s prob_s \delta_s$

4.4.5 Conclusion

By utilising this method, the trade-off between the considered objectives (Pareto curve) can be obtained not only for the nominal case, but also when there is uncertainty about some of the parameters defining the production/distribution scenario. In this case, a Pareto stochastic curve can be obtained and the comparison with the equivalent deterministic one would demonstrate the convenience of using the stochastic formulation. The effects of these uncertainties can be accounted for as a risk associated with the NPV of the investment, which had been introduced as an additional objective into the model. Then, this risk can be managed to reduce the probability of having low earnings derived from the investment. The interaction between the design objectives has been shown. This way of generating different possible configurations will help the decision-maker determine the best design according to the selected objective.

The supply chain, under uncertainty, optimisation technique utilised in this method was extracted and is summarised in Table 4-7.

4.5 Optimisation of a Supply Chain Network, under different configuration scenarios, and subject to conditions of Multi-objective Fuzzy Uncertainty

- based on the work of Tsiakis and Papageorgiou (2007) -

4.5.1 Introduction

The paper of Tsiakis and Papageorgiou (2007) focused on the determination of optimal operating conditions for a supply network operating under different network configuration scenarios, subject to multi-objective fuzzy operating conditions and also subject to certain operating and financial constraints. The production and distribution supply chain network under consideration consisted of a set of existing or potential manufacturing facilities, warehouses, distribution centres with multiple supply configurations and customer zones with varying demand. In addition to this, certain optimum operating questions need to be answered. Such questions include choice of production facility, choice of production mix per production facility, choice of which distribution centre(s) supplies which customers and then, finally, what inventory levels are necessary to maintain desired service levels.

Therefore the work of Tsiakis and Papageorgiou (2007) proposes a strategic planning framework for multi-echelon supply chain networks that integrates all components associated with production facility location as well as that of distribution centre location.

4.5.2 Description of Problem

The optimal design and operation of multi-product, multi-echelon global production and distribution networks is considered in this research work. The network consists of a number of existing multi-product manufacturing sites at fixed locations, a number of distribution centres and, finally, a number of customer zones also at fixed locations. Most products can be produced at different plants sited at different locations. Distribution Centres (DC's), which are differentiated according to the upper and lower bound limits of their material handling capacities, typically supply more than one customer zone and typical customer zones have demand for more than one type of product.

Operational costs include those costs associated with production, handling of material at distribution centres, transportation and duties. Transportation costs are modelled as linear functions of the actual flow of the product from source to destination and, additionally, taxes and cost duties are imposed on transportation costs. Optimal supply chain network operating decisions to be provided include product production portfolios per plant, product quantities to be produced, plant utilisation levels and choice of transportation links. The overall operational objective is to minimise operating costs.

4.5.3 Generation of a program model for a global production and distribution network

The global supply chain problem, described above, can be formulated as a Mixed Integer Linear Program (MILP), from which an optimal design of the network can be determined. The nomenclature for this work is described in Table 4-9. The proposed mathematical model for this MILP is described in Table 4-8 below:

Table 4-8: MILP model for Global Supply Chain Problem

Objective Function (minimisation of total costs)	Formulation
1) Plant Infrastructure Costs 2) Distribution Infrastructure Costs 3) Production Costs 4) Material Handling Costs at Distribution Centres 5) Transportation Costs 6) Duties Cost	ΣCPcjYj + ΣCPsj (1 - Yj) ΣCDcjYj + ΣCDsj (1 – Yk) ΣCPijPij + τΣijCijPNijCWij + ΣijCisOik ΣikCikDH (ΣlQikl) ΣijkCijkTOijk + ΣijkCiklTOikl ΣijkCijkDQijk + ΣiklCiklDQikl
Subject to: Type and Nature of Constraints	Formulation
1) Network structure constraints - Link between plant, *j* and DC, *k* only if plant established	$X_{jk} \leq Y_j \qquad \forall j, k$

- If DC k exists, then can be fed by more than one plant, j	$\sum X_{jk} \geq Y_k \quad \forall k$
- Link between DC, k and Cust. Zone, l only if DC, k exists	$X_{kl} \leq Y_k \quad \forall k, l$
- Each Cust. Zone may be supplied by more than one DC	$\sum X_{kl} \geq 1 \quad \forall l$
2) Logical constraints for transportation flows	
- Flow from plant, j to DC, k only if link exists	$Q^{min}_{jk} X_{jk} \leq \sum Q_{ijk} \leq Q^{max}_{jk} X_{jk}, \quad \forall i, j, k$
- Flow from DC, k to cust zone, l only if link exists	$Q^{min}_{kl} X_{kl} \leq \sum Q_{ikl} \leq Q^{max}_{kl} X_{kl}, \quad \forall i, k, l$
3) Material Balances	
- Prod. of i by plant, j must equal flow of i to all DC's, k	$P_{ij} = \sum Q_{ijk} \quad \forall i, j$
- Total flow of i from DC = total flow i entering SCN node	$\sum Q_{ijk} + O_{ik} = \sum Q_{ikl} \quad \forall i, k$
- Total flow of i to from DC, k to cust zone, l = demand, D	$\sum Q_{ikl} = D_{il} \quad \forall i, l$
4) Production Constraints	
- Production within limits	$Y_j P^{min}_{ij} \leq P_{ij} \leq Y_j P^{max}_{ij} \quad \forall i, j$
- Production restriction	$\sum T_{ij} \leq (H_j - M_j) Y_j - \tau \sum N^c_{ij} W_{ij} \quad \forall j \; T^{min}_{ij}$
- Production availability limitation	$W_{ij} \leq T_{ij} \leq T^{max}_{ij} W_{ij} \quad \forall i, j$
- Production by campaigns	$P_{ij} = r^d_j T_{ij} \quad \forall i, j$
- Plant utilisation	$U_j = \sum T_{ij} \quad \forall j$
- Uniform multi-site production distribution	$\Delta^U \geq U_j - U_{j^*} \quad \forall j, j^* \neq j$ $\Delta^U \geq U_{j^*} - U_j \quad \forall j, j \neq j^*$ $\Delta^U \leq \zeta$
- Distribution Centre capacity	$D^{min}_{ik} Y_k \leq D_k \leq D^{max}_{ik} Y_k \quad \forall k$ $D_k \geq \sum \delta_{ik} Q_{ikl} \quad \forall k$

Table 4-9: Nomenclature for Multi-objective Fuzzy Modelling - Tsiakis and Papageorgiou (2007)

Index/Set	Physical Meaning
I	Products
J	Plants
K	Possible distribution centres
L	Customer demand zone
Parameters	Physical Meaning
$C^{DH}{}_{ik}$	unit handling cost for product *i* at distr. centre *k* centre at loc. *K*
$C^{Dc}{}_{k}$	annual fixed cost of establishing down a distribution centre at location *k*
$C^{Ds}{}_{k}$	annual fixed cost of shutting down a distribution centre at location *k*
$C^{d}{}_{ijk}$	unit duty cost for prod *i* leaving production plant *j* to Distribution centre *k*
$C^{d}{}_{ikl}$	unit duty cost for product *i* leaving distribution centre *k* to customer *j*
$C^{Pc}{}_{j}$	annualised fixed cost of establishing a production plant at location *j*
$C^{Ps}{}_{j}$	annualised fixed cost of shutting down a production plant at location *j*
$C^{P}{}_{ij}$	unit production cost for product *i* at plant *j*
$C^{S}{}_{i}$	unit production cost for product *i* supplied to the network from third parties
$C^{T}{}_{ijk}$	unit transport cost for product *i* from plant *j* to distribution centre *k*
$C^{T}{}_{ikl}$	unit transport cost for product *i* from distribution centre *k* to customer zone *l*
$D^{min}{}_{k}/D^{max}{}_{k}$	minimum/maximum distribution centre capacity
D_{il}	demand for product *i* in customer zone *l*
H_j	operating horizon of production site *j* in days per annum
M_j	annual days of maintenance of production site *j*
$N^{C}{}_{ij}$	number of campaigns of product *i* at production plant *j*
$P^{min}{}_{ij}$ / $P^{max}{}_{ij}$	minimum/maximum production capacity of plant *j* for
$Q^{min}{}_{jk}$ / $Q^{max}{}_{jk}$	product *i*
$Q^{min}{}_{lk}$ / $Q^{max}{}_{lk}$	minimum/maximum rate of flow of materials transferred
$T^{min}{}_{ij}$ / $T^{max}{}_{ij}$	from plant *j* to distribution centre *k*
	minimum/maximum rate of flow of materials transferred from distribution centre *k* to customer zone *l*
$r^{d}{}_{j}$	minimum/maximum availability of plant *j* for the
α_{jk}	production of product *i* in days

	expected daily production rate of plant *j*
α_{kl}	duty coefficient relating production plant *j* to distribution
β	centre *k*
γ	duty coefficient relating distribution centre *k* to customer
δ_{ik}	zone *l*
ζ	coefficient applied to production cost for duty purposes
τ	coefficient applied to transportation cost for duty purposes
	coefficient relating capacity of distribution centre *k* to flow of product *i* handled
	utilisation parameter in days
	change-over coefficient in days
Continuous Variable	Physical Meaning
D_k	capacity of distribution centre *k*
O_{ik}	out-sourced product *i* delivered to distribution centre *k*
P_{ij}	production rate of product *i* in plant *j*
Q_{ijk}	rate of flow of product i transfer red from plant *j* to distribution centre *k*
Q_{ikl}	rate of flow of product *i* transferred from distribution centre *k* to customer zone *l*
T_{ij}	days allocated for the production of product *i* in plant *j*
U_j	utilisation of production plant *j* as production days per year
Δ^U	maximum allowed difference in utilisation between plants
Binary Var.	Physical Meaning
Y_j	1 if production plant *j* is to be established, 0 otherwise
Y_k	1 if distribution centre *k* is to be established, 0 otherwise
X_{jk}	1 if a production plant *j* is to be assigned to a distribution centre *k*, 0 otherwise
X_{kl}	1 if a distribution centre *k* is to be assigned to a customer zone *l*, 0 otherwise *k*, 0 otherwise
W_{ij}	1 if production plant *j* is to produce product *i*, 0 otherwise

4.5.4 Case Study

To demonstrate the applicability of the proposed model, the performance of a global organisation with a supply chain network extending in many countries, where multiple currencies and taxes apply, is examined.

The model was applied to a network consisting of six manufacturing plants located in different countries on different continents. Each plant can produce six types of products using a number of shared production resources, therefore only one product may be produced at a time. A number of possible distribution centres are considered to handle the material based on the market demand placed by the customer zones Customer zones comprise eight large geographical areas. The network superstructure is given in Fig 4-4. The superstructure assumes complete connectivity between all sites, which is subject to constraints regarding material transfers.

Different scenarios are examined to demonstrate the benefits of optimisation over decisions imposed by the practice of heuristics. These scenarios are:

- free optimisation where all decisions are determined by the model subject to the constraints
- optimise network flows when product allocation decisions and network structure has been fixed
- optimise the network when each plant can manufacture up to three allocated products
- examine the network sensitivity by changing some of the parameters such as utilisation constraints

Problem Description

Production Plants – The theoretical maximum production capacity of each plant, i.e. P_{ij}^{max}, with respect to each product is given in Table 4-10 and the corresponding minimum production rates, P_{ij}^{min}, are taken to be zero. The operating horizon is 365 days per plant and maintenance duration is 10 days, except for plants 5 and 6, which are 7 days.

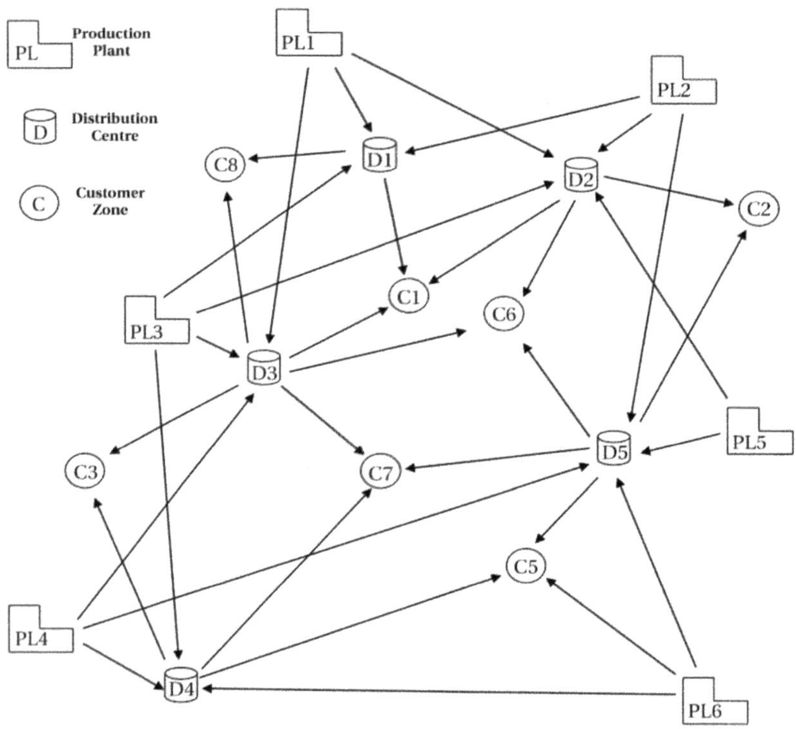

Figure 4-4: Typical Supply Chain Representation

a) Production plants - Maximum production time for each product per site is $T_{ij}^{max} = 365$ days, whilst the minimum campaign length is $T_{ij}^{min} = 7$ days. The change-over duration is assumed to be $\tau = 1$ day and it is assumed that the utilisation coefficient is $\zeta = 10$ days. The unit production costs per product for each plant are given in Table 4-11

Table 4-10: Daily production rate per production site (tons/day)

P1	P2	P3	P4	P5	P6
73.4	12.7	35.1	34.4	11.0	10.0

Table 4-11: Unit production cost per plant (rmu/day)

Product Site	PL1	PL2	PL3	PL4	PL5	PL6
P1	4.68	4.53	5.17	5.04	4.27	4.26
P2	3.23	3.17	3.21	3.87	3.12	3.97
P3	4.18	4.19	5.17	4.35	3.71	3.70
P4	2.48	2.71	2.54	2.74	2.43	2.42
P5	1.43	1.49	1.55	1.41	1.28	1.27
P6	2.15	2.15	2.22	2.40	2.07	2.06

Table 4-12: Transportation cost between production sites and distribution centres

DC \ Prod Site	D1	D2	D3	D4	D5	D6
PL1	0.000	0.062	0.081	0.053	0.051	0.056
PL2	0.065	0.089	0.127	0.077	0.071	0.077
PL3	0.083	0.126	0.000	0.141	0.125	0.119
PL4	0.107	0.138	0.182	0.000	0.143	0.144
PL5	0.183	0.236	0.117	0.038	0.000	0.038
PL6	0.168	0.178	0.236	0.117	0.038	0.000

Table 4-13: Values of coefficient, α_{jk}, to apply to duties structure

DC \ Prod Site	Prod Group	D1	D2	D3	D4	D5	D6
PL1	P1-P6	0.0	0.0	7.1	53.1	10.0	5.0
	P5	0.0	0.0	3.8	53.1	10.0	12.5
PL2	P1-P6	0.0	0.0	7.1	53.1	10.0	5.0
	P5	0.0	0.0	3.8	53.1	10.0	12.5
PL3	P1-P6	6.5	9.0	0.0	53.1	10.0	5.0
	P5	6.5	9.0	0.0	53.1	10.0	12.5
PL4	P1-P6	0.0	9.0	4.1	0.0	10.0	5.0
	P5	0.0	9.0	0.3	0.0	10.0	12.5
PL5	P1-P6	0.0	9.0	7.1	53.1	0.0	5.0
	P5	0.0	9.0	3.8	53.1	0.0	12.5
PL6	P1-P6	6.5	9.0	4.1	53.1	10.0	0.0
	P5	6.5	9.0	0.3	53.1	10.0	0.0

Table 4-14: Costs associated with the distribution centres upon establishment

DC \ Prod Site	D1	D2	D3	D4	D5	D6
Fixed Infrastructure costs (,000 rmu)						
	4,300	2,900	3,000	2,200	1,300	1,500
Material handling costs per product (rmu/kg)						
P1	0.070	0.068	0.077	0.076	0.064	0.064
P2	0.048	0.048	0.048	0.058	0.047	0.060
P3	0.063	0.063	0.077	0.065	0.056	0.056
P4	0.037	0.041	0.038	0.041	0.036	0.036
P5	0.021	0.022	0.024	0.021	0.009	0.019
P6	0.032	0.032	0.033	0.036	0.034	0.031

Table 4-15 Customer zone demand per product (tons)

Cust Zone \ Product	C1	C2	C3	C4	C5	C6	C7	C8
P1	5,701	3,665	3,398	2,686	1,342	1,819	2,436	1,809
P2	2,116	365	1,468	1,355	910	748	1,049	748
P3	3,246	887	1,369	1,353	609	679	1,209	679
P4	2,582	295	3,016	466	556	906	1,436	906
P5	3,683	437	2,284	922	794	844	1,578	844
P6	294	64	311	1,173	130	323	172	323

Table 4-16: Transportation cost between distribution centre and customers (in rmu/kg)

Cust Zone \ DC	C1	C2	C3	C4	C5	C6	C7	C8
DC1	0.00	0.62	0.81	0.53	0.58	0.56	1.31	1.66
DC2	0.65	0.00	1.27	0.77	0.73	0.77	1.38	1.77
DC3	0.90	1.40	0.00	1.41	1.25	1.19	1.35	1.29
DC4	1.13	1.52	1.82	0.00	1.43	1.44	1.93	1.54
DC5	1.89	2.16	2.61	1.66	0.00	0.57	1.57	1.34
DC6	1.74	1.78	2.36	1.17	0.00	0.59	1.36	1.23

Table 4-17: Duty coefficient for material transferred between distribution centre and customer αkl(%)

Cust Zone \ DC	C1	C2	C3	C4	C5	C6	C7	C8
DC1	0.0	0.00	7.1	53.1	12.5	11.0	17.0	15.0
DC2	0.0	0.00	7.1	53.1	12.5	11.0	17.0	15.0
DC3	6.5	9.0	0.0	53.1	12.5	11.0	17.0	15.0
DC4	6.5	9.0	4.1	0.0	12.5	11.0	17.0	15.0
DC5	6.5	9.0	7.1	53.1	0.0	11.0	17.0	15.0
DC6	6.5	9.0	4.1	53.1	12.5	11.0	17.0	15.0

b) Distribution Centres – The distribution centres required are assumed to have sufficient capacity to handle the material received and distributed. In order for a connection between a production plant and a distribution centre to exist, there must be at least a total flow of 1,000 tons, i.e. ($Q_k^{min} \geq 1{,}000$ tons $\forall j$), Unit transportation costs, which are independent of the product type and the quantity transported, from the production plants to the distribution centres are given in Table 4-10,. The quantities transported to the distribution centres are subject to duties, as previously defined and Table 4-13 describes the coefficients used between production sites and distribution centres. The establishment costs of distribution centres are treated as fixed costs for the organisation and are amortised as annualised costs and, depending on the type of material stored, are treated as a handling cost per product. Such costs are given in Table 4-14.

c) Customer Zones – Customer zones are situated in close proximity to distribution centres. The demand per product per customer zone is given in Table 4-15. The transportation costs from the distribution centres to customer demand zones are given in Table 4-16 and the amounts transported to customer demand zones are subject to duties. Table 4-17 describes the duty coefficients used between distribution centres and customer zones.

Case 1: Network Optimisation – Base Scenario

The first case is a free optimisation exercise where the model is allowed to determine all the decisions required such as product allocation per plant and production level, establishment of distribution centres and customer allocation, material flows and distribution centre capacity. The optimal solution for the network is given in Fig 4-5.

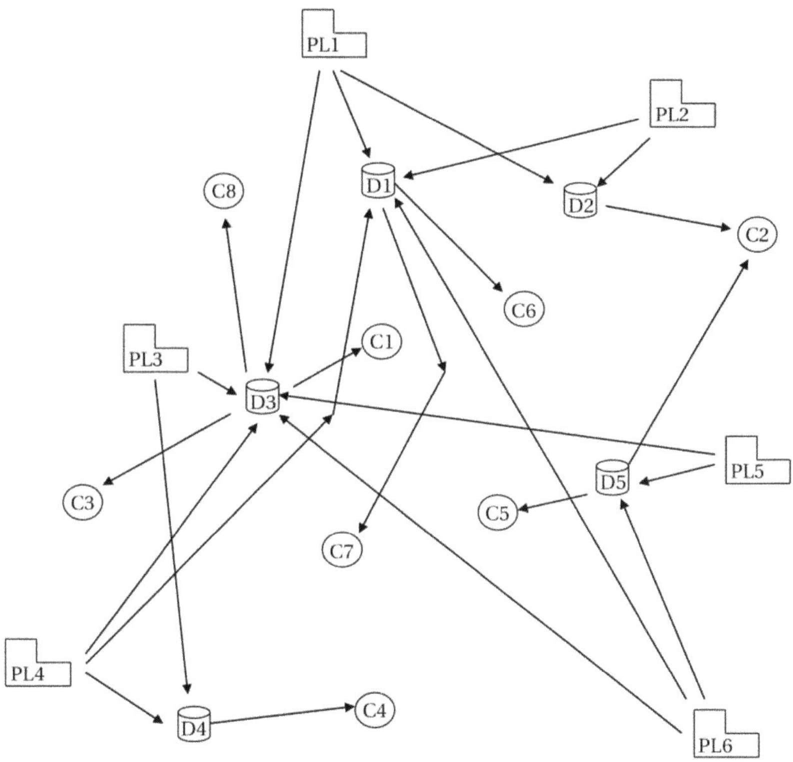

Figure 4-5: Optimal network configuration for base case

Table 4-18 shows the product allocation per plant and the number of day's production per product. Outsourced material is purchased to cover demand that cannot be satisfied by internal production. The material is directly supplied to the distribution centres and from there to the assigned customer zones.

The total material supplied from the production plants to distribution centres is shown in Table 4-19. This is subject to the material flow constraint that applies for an established link.

Table 4-18: Product allocation per plant and production days allocated

Product Act Prod	PL1	PL2	PL3	PL4	PL5	PL6	Out-Sourced
P1	148	321	0	173	131	133	0
P2	54	0	61	42	108	0	0
P3	69	0	50	66	0	58	0
P4	0	0	124	0	0	48	5097
P5	58	0	95	0	0	94	0
P6	0	14	0	50	96	0	2289
Utilisation Total	329	335	330	331	335	333	

Material flows from a distribution centre, where it has been established for the assigned customers, and is equal to the total demand placed by the customer. The total annual operational cost for this operation is 311,681,000 rmu.

Table 4-19: Material flow from production plants to distribution centres (tons)

From/To Prod Site Dist Centre	DC1	DC2	DC3	DC4	DC5	DC6
P1	20,765	2,096	1,000	0	0	0
P2	1,000	3,258	0	0	0	0
P3	1,000	0	10,538	0	0	0
P4	1,418	0	3,656	6,316	0	0
P5	1,000	0	0	0	2,340	0
P6	1,000	0	1,337	0	1,315	0

Case 2: Fixed product allocation and customer assignment

This case assumes that production allocation and network configuration, in terms of customer allocation to distribution centres, are decisions already taken. Product allocation is done according to plant capability and then customer demands are assigned to the nearest possible geographical site. As a result of this, the model size is reduced in terms of the number of the decision variables that need to be determined since X_{kl} and W_{ij} are fixed. Production levels are then optimised, in terms of days per year and number of campaigns, and flows within the network as well as connections between plants and distribution centres. Table 4-20 shows the activity levels for each plant based on the pre-allocation of products. Out-sourced material purchased for this case is given in the last column of Table 4-20. More products are out-sourced and its total is higher than that of the base case.

Table 4-20: Production per plant (tons) as days per product

Product Act Prod	PL1	PL2	PL3	PL4	PL5	PL6	Out-Sourced
P1	91	327	84	165	69	133	1,711
P2	50	0	60	42	91	42	0
P3	62	0	42	70	59	58	134
P4	42	0	56	0	28	28	4,323
P5	74	0	75	49	80	66	0
P6	0	0	0	0	0	0	2,467
Utilisation Total	319	327	317	326	327	327	

The flow of material between plants and distribution centres is given in Table 4-22 and the total cost of the operation, in this case, is 327,049,000 rmu. The breakdown of individual costs is given in Table 4-21 and the solution of the problem is shown in Fig 4-6, which also shows the customer assignment to distribution centres. These connections were fixed

Table 4-21: Material flow between production plants and distribution centres

From/To Prod Site Dist Centre	DC1	DC2	DC3	DC4	DC5	DC6
PL1	19,869	2,197	0	0	0	1,353
PL2	1,000	3,156	0	0	0	0
PL3	0	0	9,482	0	0	1,634
PL4	1,000	0	1,000	6,316	0	2,936
PL5	0	0	0	0	3,260	0
PL6	0	0	0	0	0	3,583
Total handling capacity	22,362	5,713	11,846	7,955	4,341	13,209

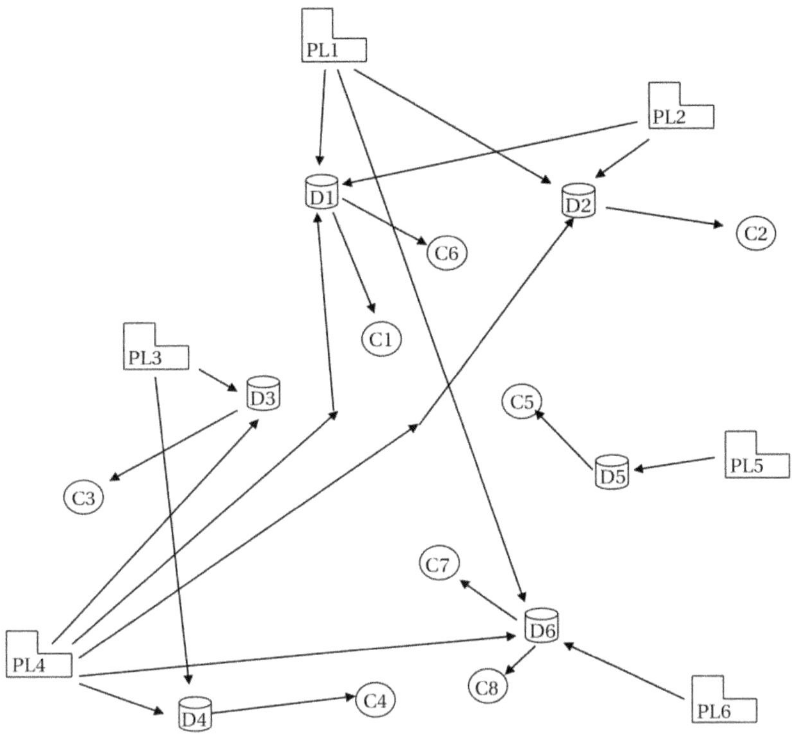

Figure 4-6: Optimal network configuration for fixed product allocation and customer assignment

Case 3: Maximum allocation of three products per plant

In this case it is assumed that each plant is allowed to manufacture up to three products. All plants can produce a whole range of products and optimisation is used to determine the product allocation, material flows within the network and the connections between plants, distribution centres and customer zones.

The actual production per site, in terms of days, is given in Table 4-22 and the flow of material in the network is given in Table 4-23. Fig 4-7 shows the optimal network configuration and the customer allocation to distribution centres, and the associated cost is 343,218,000 rmu and the cost analysis is displayed in Table 4-25.

Table 4-22: Production levels in days per plant

Plant\ Product	PL1	PL2	PL3	PL4	PL5	PL6	Out-Sourced
P1	131	283	84	145	118	98	0
P2	0	0	0	74	109	107	3,840
P3	82	0	0	70	0	0	1,344
P4	0	0	125	0	0	0	5,574
P5	86	0	90	0	72	94	0
P6	0	16	0	0	0	0	2,269
Utilisation Total	289	289	289	289	289	289	

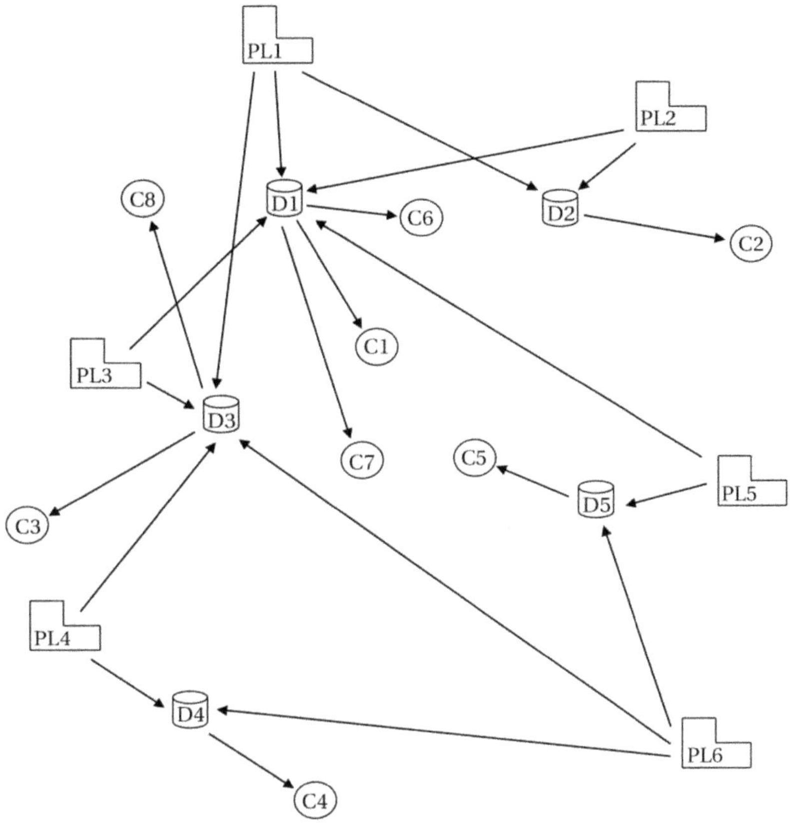

Figure 4-7: Optimal network configuration where product allocation has been fixed

Table 4-23: Material flow between production plants and distribution centres

DC\Plant	DC1	DC2	DC3	DC4	DC5	DC6
PL1	19,869	2,197	0	0	0	1,353
PL2	1,000	3,156	0	0	0	0
PL3	0	0	9,482	0	0	1,634
PL4	1,000	0	1,000	6,316	0	2,936
PL5	0	0	0	0	3,260	0
PL6	0	0	0	0	0	3,583
Total handling capacity	22,362	5,713	11,846	7,955	4,341	13,209

Case 4: Sensitivity Analysis on utilisation factor and change-over duration

The base network optimisation scenario was resolved at different values of utilisation factor, ζ. As a result of these changes, no modification to the solution of the base case was observed. This primarily has to do with the fact that there is no spare capacity available in the production plants and therefore they are utilised to the maximum available rate.

Process change-overs have been identified, by many companies, as one of the major factors limiting the production capacity and flexibility of the plants. The case of change-over duration being halved to 12hrs, as opposed to 1 day, will now be examined. A reduction in change-over duration results in changes to product portfolio per plant, which is summarised in Table 4-24. Out-sourced material is purchased to cover demand that cannot be satisfied by internal production and the amount of material that is consequently purchased is smaller than that bought in the base case because of increased utilisation. The network flow of material is given in Table 4-24 and the total cost of operation, in this case, is 295,714,000 rmu and a cost analysis is given in Table 4-25. The network structure remains the same as that in the base case, i.e. Fig 4-5.

Table 4-24: Production levels in days per plant

Plant\ Product	PL1	PL2	PL3	PL4	PL5	PL6	Out-Sourced
P1	128	334	84	116	131	133	0
P2	48	0	60	42	109	42	0
P3	82	0	0	70	59	58	122
P4	0	0	120	0	49	28	4957
P5	82	0	75	50	0	84	0
P6	0	14	0	62	0	0	130
Utilisation Total	340	348	339	289	347	345	

Results Analysis

The results obtained demonstrate clearly the savings of optimisation over heuristic decision-making in supply chain network structure design. The benefits are manifested in an improved utilisation of plants by means of preferred portfolio selections, minimisation of out-sourced material and savings on transportation and duties. Although the plants are multi-purpose, it is preferable to operate specific plants in a dedicated mode. It is also clear that further savings in operational costs are possible from a reduction in change-overs as well as from a reduction in the purchase of out-sourced material. A cost comparison analysis between all four cases is shown in Table 4-25

Table 4-25: Cost analysis and comparison of the various cases

Cost analysis \ Expense category	Case 1	Case 2	Case 3	Case 4
Global production costs	225,641	223,558	206,798	222,863
Global distribution costs	22,060	21,397	21,923	21,979
Global duties paid	4,503	5,942	5,197	4,392
Global out-sourcing costs	42,230	57,548	92,053	29,233
Infrastructure costs	13,800	15,300	13,800	13,800
Material handling costs	3,447	3,394	3,447	3,447
Total Costs	341,681	327,049	343,118	295,714
Difference (%)	0.00	+4.93	+10.08	-5.12

4.5.5 Extraction of the 'Optimisation of a Supply Chain Network under Multi-objective Fuzzy Uncertainty' Technique

Table 4-26: Optimisation under Multi-objective Fuzzy Uncertainty Technique

1	Describe the supply chain network environment problem including an operational summary of all subcomponents, i.e. production centres, distribution centres and customer zones
2	Describe the ideal operational requirements of the system, e.g. product portfolio and utilisation level per production plant and the minimisation of total annualised costs of the network
3	Specify the notation, i.e. indices, parameters and variables, to be employed in the formulation of a MILP (mixed integer linear programming) model
4	Mathematically define all constraints of the problem - Network structure constraints - Logical constraints for transportation flows - Material balances - Production constraints
5	Mathematically define the objectives of the problem, i.e. the minimisation of:: - Fixed infrastructure costs - Production costs - Material handling costs at distribution centres - Transportation costs - Duties
6	Solution of the supply chain network MILP model via a number of optimisation scenarios using MS-Excel Solver: - Free optimisation exercise in which the model is allowed to determine all the required decisions - Fixed product allocation and customer assignment - Maximum allocation of products per plant
5	Results analysis
6	Conclusion

4.5.6 Conclusion

This paper proposed an integrated model based on a detailed mathematical programming formulation that addressed some of the complex issues related to the design and operation of global supply chain networks. The focus was on certain business benefits such as the operational and distribution efficiency of the network, operational cost, customer satisfaction and product quality.

4.6 Optimisation of a Supply Chain Network under Multi-objective Stochastic Fuzzy Uncertainty

- based on the work of Chen et al. (2007) –

4.6.1 Introduction

In a paper by Chen et al (2007), the researchers decided to extend previous supply chain, under uncertainty, research work of authors such as Gupta and Maranas (2003) and Tsiakis and Papageorgiou (2006) by adopting a scenario-based approach to handle market demand uncertainty, with each scenario having a known probability. The intention behind the authors' work was to determine optimal infrastructural considerations such as the number, location and capacities of warehouses and distribution centres as well as to determine actual optimal operating conditions such as production plans, costs and logistical considerations for a supply chain network operating under uncertainty.

A supply chain consists of a number of integrated process operations whereby a number of different business entities, such as suppliers, producers, distributors and retailers, work and cooperate together to acquire raw materials, convert them into final products, distribute them to various wholesalers who then supply various retailers who, in turn, supply different customer zones.

4.6.2 Problem Description

The supply chain network under consideration consists of a number of multi-product production plants, candidate warehouses and distribution centres and a number of customer zones as depicted in Fig 4-8. A number of assumptions are made with regard to modelling and optimisation. Firstly, the whole system is operated steadily; with neither stock accumulation nor stock depletion and therefore the possibility of inventory build-up can be ignored.

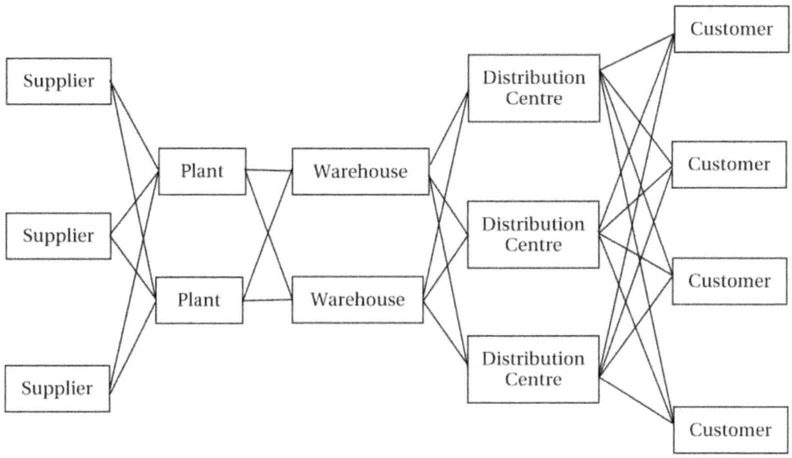

Figure 4-8: ***A representation of the studied supply chain network***

4.6.3 Supply Chain, under Demand Uncertainty, modelling

An integrated multi-echelon supply chain model is considered for optimal decisions-making and a scenario-based representation for uncertain product demands is considered for uncertain market demand modelling. The indices, sets and parameters used for modelling the supply chain network with product demand uncertainty are given in the nomenclature in Table 4-27. Parameters are divided into two categories: firstly, there are cost parameters, including product cost, handling cost and transport cost; and, secondly, there are other parameters that describe system information criteria such as material handling, transport capacity and forecasting of customer demand. Additionally, two kinds of variables are employed in the model, firstly there are binary variables that represent policy decisions in determining the viability of establishing warehouses and distribution centres, along with using economies of scale for manufacturing or transportation and, secondly, there are continuous variables that describe manufacturing quantities, handling capacities and transport quantities.

Such model is based on the work of Chen et al (2007).

Table 4-27: Nomenclature for Multi-objective Fuzzy Uncertainty Modelling – Chen et al. (2007)

Index/Set	Physical Meaning
$c \in C$	Customer Zone
$d \in D$	Distribution Centres
$i \in I$	Products
$k \in K$	Transport Capacity Level
$m \in M$	All Objectives
$p \in P$	Plants
$s \in S$	Scenarios
$t \in T$	Periods
$w \in W$	Warehouses
Parameters	**Physical Meaning**
FCD^{i}	Forecasting customer demand *i* for customer *c*
FTC^{k}_{*}	k^{th} level fix transport cost, *p to w, w to d, d to c*
LL_{*}	Local incentive of *w, d*
PPD_{s}	Probability of product scenario *s*
PQ^{*}_{ips}	Maximum, minimum manufacturing quantity of product *i*
$Q^{max}{}_{*s}$	Maximum transport quantity of *p* to *w*, *w* to *d*, *d* to *c*
$Q^{min}{}_{*s}$	Minimum transport quantity of *p* to *w*, *w* to *d*, *d* to *c*
R_{*nts}	Total resource *n* at *p*
$SQ^{+}{}_{*}$	Maximum, minimum capacity of *w*, *d*, + ∈ [max, min}
TCL^{k}_{*}	k^{th} transport capacity level, *p* to *w*, *w* to *d*, *d* to *c*
TT_{*}	Transport time of *p* to *w*, *w* to *d*, *d* to *c*
UEC_{*}	Unit establishing cost of *w*, *d*
UHC^{i}_{*}	Unit handling cost of product *i* for *w*, *d*
UPC^{i}_{*}	Unit production cost of product *i* for plant *p*
UTC^{k}_{*}	k^{th} level unit transport cost, *p* to *w*, *w* to *d*, *d* to *c*
$\alpha^{i}{}_{*}$	coefficient relating the capacity of *d* to flow of product *i* handled
$\beta^{i}{}_{*}$	coefficient relating the capacity of *w* to flow of product *i* handled
$\rho^{i}{}_{*nts}$	coefficient for resource *m* used in plant *p* for product *i*
Variable	**Physical Meaning**
J_{*}	Objectives
OTT	Total transportation time
$PQ^{i}{}_{*ts}$	Manufacture quantity of *i*
$Q^{i}{}_{*ts}$	Total transport quantity, *p* to *w*, *w* to *d*, *d* to *c*
SQ_{*}	Capacity of *w*, *d*
TCO	Total Cost
TEC	Total estab. cost of all warehouses and distribution centres

THC_{ts}	Total handling cost of scenario *s* in period *t*
TLI, LD	Total local incentive and overall index on distribution centres
TPC_{ts}	Total production cost of scenario *s* in period *t*
TTC_{ts}	Total transportation cost of scenario *s* in period *t*
TTC_{*ts}	Total transportation cost of * for scenario *s* in period *t*
TTQ_{*ts}	k^{th} level transport quantity, *p* to *w*, *w* to *d*, *d* to *c*
Binary Var.	Physical Meaning
X_*	A link between *p* and *w*, *w* and *d*, *d* and *c* exists
Y_*	Warehouse *w* or distribution centre *d* is to be established
Z^k_{*ts}	k^{th} transport capacity level, *p* to *w*, *w* to *d*, *d* to *c*
Fuzzy Var.	Physical Meaning
FD	Fuzzy set for final decision
τ_m	Fuzzy set for objective *m*

The scope of the modelling process will encompass all those categories usually associated with a supply chain network, i.e.:

i) Network structure constraints

ii) Transportation constraints

iii) Material Balance constraints

iv) Production resource constraints

v) Warehouse and Distribution Centre capacity constraints

vi) Overall Cost constraints

a) Network structure constraints

There are network infrastructural constraints between all plants, warehouses, DC's and customer zones, which may be summarised as follows:

$$X_{pw} \leq Y_w; \qquad X_{wd} \leq Y_w \qquad (1)$$

Eq. (1) recognises that a link between a plant *p* and a warehouse *w* exists, or that a link between a warehouse *w* and a DC (Distribution Centre) *d* can only exist if a warehouse *w* exists.

$$X_{wd} \leq Y_d; \qquad X_{wd} \leq Y_d \qquad (2)$$

Eq. (2) expresses a link between a DC (Distribution Centre) *d* and a warehouse *w*, but only if the DC exists.

It is also assumed that a DC (Distribution Centre) can only be served by a single warehouse and that a customer zone can only be served by a single DC, as shown in Eq. (3).

$$\Sigma_{\forall\, w \in W}\, X_{wd} = Y_d; \; \Sigma_{\forall\, d \in D}\, X_{dc} = 1 \qquad (3)$$

$$\forall\; p \in P,\; w \in W,\; d \in D,\; c \in C$$

b) Transportation constraints

The transportation constraints at considered periods, $t \in T$, in different economic scales are given below:

$$TCL_{*}^{k-1} Z_{*ts}^{k} < TTO_{*ts}^{k} \leq TCL_{*}^{k} < TCL_{*}^{k} Z_{*ts}^{k} \quad (4)$$

$$\Sigma_{\forall k \in K} Z_{*ts}^{k} \leq 1 \quad (5)$$

$$\Sigma_{\forall i \in I} Q_{*ts}^{\mathbf{i}} = \Sigma_{\forall k \in K} Q_{*ts}^{\mathbf{k}} \quad (6)$$

$$Q_{*s}^{min} X_{*} \leq \Sigma_{\forall k \in K} Q_{*ts}^{\mathbf{i}} < Q_{*s}^{max} X_{*} \quad (7)$$

$$Q_{*ts}^{\mathbf{i}} \geq 0 \quad (8)$$

Where $* \in \{pw, wd, dc\}$

$\forall\ p \in P, w \in W, d \in D, c \in C, i \in I, t \in T, s \in S$

Eqs. (4) and (5) reflect the fact that several transport capacity levels, with various associated unit transport costs, can be used for a three-tier supply chain network, as depicted in Fig 4-7 and it is further assumed that no more than one transport capacity level can be chosen per period. Eq. (6) reflects the fact that singular transport quantities from plants to warehouses, from warehouses to DC's or from DC's to consumer zones, per period, are translated respectively into overall transport quantities. Eq. (7) recognises the fact that total transport quantities have lower and upper bound limits for all existing links.

c) Material balance constraints

It is assumed that the operation is in a constant steady state, therefore there is no stock accumulation or depletion. All material balance constraints can therefore be summarised as follows:

$$PQ_{pts}^{i} = \Sigma_{\forall w \in W} Q_{pwts}^{i} \quad (9)$$

$$\Sigma_{\forall p \in P} Q_{pwts}^{i} = \Sigma_{\forall d \in D} Q_{dwts}^{i} \quad (10)$$

$$\Sigma_{\forall w \in W} Q_{dwts}^{i} = \Sigma_{\forall c \in C} Q_{dcts}^{i} \quad (11)$$

$$\Sigma_{\forall d \in D} Q_{dcts}^{i} = FCD_{cts}^{i} \quad (12)$$

$$PQ_{ips}^{t} \geq 0 \quad (13)$$

$\forall\ p \in P, i \in I, t \in T, s \in S$

Eq. (9) states that the production of a product i by plant p must be equal to the total flow of the product i to all warehouses. Eq. (10) states that the total flow of the product i from all plants to a warehouse w must be equal to the total flow of the product i from the warehouse to all DC's.

Similarly, Eq. (11) reflects the fact that total flow of the product i from all warehouses to a DC d must equal the total flow of the product i from the DC d to all customer zones. Also, eq. (10) reflects the fact that total flow of a product i from all DC's to a customer zone c must equal the forecasted demands.

d) Production resource constraints

It must be ensured in network design that production plants must be able to satisfy the demands of the customers. The production of each product at any plant has therefore certain limitations:

$$PQ_{ips}^{min} \leq PQ_{pts}^{i} \leq PQ_{ips}^{max} \quad (14)$$

$$\Sigma_{i \in I} \rho_{pnts}^{i} PQ_{pts}^{i} \leq R_{pnts} \quad (15)$$

$\forall\ p \in P, n \in N, i \in I, t \in T, s \in S$

Here, Eq. (14) states that the plants have maximum and minimum production capacities. Some plants may be required to produce different products at different production stages, but utilising similar resources (equipment, utilities, manpower etc.). The limitations of resource utilisation can therefore be seen in Eq. (15)

e) Warehouse and Distribution Centre capacity constraints

All capacity constraints for warehouses and DC's are formulated below:

$$SQ_{w}^{min} Y_{w} \leq SQ_{w} \leq SQ_{w}^{max} Y_{w} \quad (16)$$

$$SQ_{d}^{min} Y_{d} \leq SQ_{d} \leq SQ_{d}^{max} Y_{d} \quad (17)$$

$$SQ_{w} \geq \Sigma_{\forall i \in I} \Sigma_{d \in D} \alpha_{w}^{i} Q_{wdts}^{i} \quad (18)$$

$$SQ_{d} \geq \Sigma_{\forall i \in I} \Sigma_{c \in C} \beta_{d}^{i} Q_{dcts}^{i} \quad (19)$$

$\forall\ w \in W, d \in D, c \in C, i \in I, t \in T, s \in S$

Eq. (16) states that the capacity of a warehouse *w* has lower, SQ_{w}^{min}, and upper, SQ_{w}^{max}, bounds if such warehouse is established. Similarly, constraints also apply to the capacities of the distribution centres, as shown in Eq. (17). It is also assumed that the capacities of the warehouses and distribution centres are linearly related to the materials that they handle, as expressed in Eqs. (18) and (19).

f) Overall cost constraints

The establishment costs of warehouses and distribution centres, the production costs and all material handling and transportation costs are given below:

$$TEC = \Sigma_{\forall w \in W} UEC_w Y_w + \Sigma_{\forall d \in D} UEC_d Y_d \quad (20)$$

$$TPC_{ts} = \Sigma_{\forall i \in I} \Sigma_{\forall p \in P} UPC_p^i PQ_{pts}^i \quad (21)$$

$$THC_{ts} = \Sigma_{\forall i \in I} \Sigma_{\forall w \in W} UHC_w^i (\Sigma_{\forall p \in P} Q_{pwts}^i) + \Sigma_{\forall i \in I} \Sigma_{\forall d \in D} UHC_d^i (\Sigma_{\forall p \in P} Q_{wdts}^i) \quad (22)$$

$$TTC_{*ts} = \Sigma_{\forall k \in K} (FTC_*^k Z_{wts}^k + UTC_*^k TTC_{wts}^k) \quad (23)$$

$$TTC_{ts} = \Sigma_{\forall p \in P} \Sigma_{\forall w \in W} TTC_{pws}^t + \Sigma_{\forall w \in W} \Sigma_{\forall d \in D} TTC_{wds}^t + \Sigma_{\forall d \in D} \Sigma_{\forall c \in C} TTC_{dcs}^t \quad (24)$$

Where $* \in \{pw, wd, dc\}, \forall\ p \in P, w \in W, d \in D, c \in C, i \in I, t \in T, s \in S$

Eq. (20) shows the establishment costs of warehouses and distribution centres at prospective locations. In Eq. (21), total production costs are represented by the summations of the individual production quantities of product i multiplied by the corresponding unit production costs UPC_{ip}. Eq. (22) reflects the fact that total material handling costs can be expressed as a linear function of each product being handled at warehouses and distribution centres. Eq. (23) represents transportation costs for the plant and DC respectively. In this case, transportation cost is an amalgamation of transport level-dependent fixed cost and transport quantity-dependent carrying cost. This would result in an irregular, stage-wise linear transport cost as illustrated in Fig 4-9

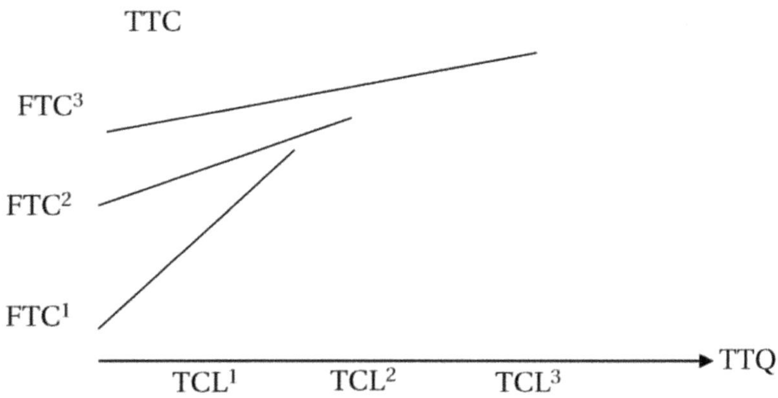

Figure 4-9 Piecewise linear relation between transport cost, TTC, and shipment quantity, TTQ

g) Optimal planning incorporating multiple objectives

For the purpose of supply chain network design, several conflicting objectives may be considered simultaneously in order to achieve an optimal design. These include the minimising the total cost, the maximising the robustness of selected objectives to demand uncertainties, the maximising the local incentives and the minimising of total transport time, and are stated below:

1) *Objective 1: Minimising the total cost*

The total cost is a summation of the total establishment costs, the total production costs, the total handling costs and the total transportation costs, and each is weighted by the respective probability of related scenario:

$$\min_{x \in \Omega} TCO = TEC + \sum_{\forall t \in T} \sum_{\forall s \in S} PPD_s(TPC_{ts} + THC_{ts} + TTC_{ts})$$

where Ω is the feasible searching space, which is a composite of all constraints, Eqns. (1) – (24), and x denotes the decision vector

$x = \{Y_w, Y_d, X_*, Q_{*ts}^{i}, TTQ_{*ts}^{k}, Z_{*ts}^{k}, PQ_{ips}^{i}, SQ_w, SQ_d\ * \in \{pw, wd, dc\}; \forall i \in I, p \in P, w \in W, d \in D, c \in C, k \in K, t \in T, s \in S\}$

2) Objective 2: Maximising the robustness to various scenarios

As has been stated previously, the application of an explicit scenario-based approach for modelling the uncertain product demands would result in the fact that all operating variables become scenario-dependant. However, as a result of this, the total cost realisation might be unacceptably high for certain scenarios, with especially low probabilities. Consequently, it would therefore be significant to reduce the variability of objective values J_s for any realisation of scenarios. An important issue in enforcing the robustness of uncertainties is the choice of variability metrics. For the total cost to be minimised, the decision maker usually does not care if the objective value J_s is lower than the expected mean value J. Consequently, the upper partial mean (UPM) is used as the measure of robustness where only costs above the expectation are penalised and are also weighted accordingly by the probabilities of related scenarios, PPD_S

$$UPM = \Sigma_{\forall s \in S} PPD_S \max\{0, J_S - J\}$$

As the upper partial mean decreases, the robustness of total cost will increase, and therefore the robustness index (RI) can be defined as below.

$$Max_{x \in \Omega}(RI) = -UPM = -\Sigma_{\forall s \in S} PPD_S UPM_S$$

$$\text{where } UPM_S \geq 0,\ UPM_S \geq J_S - J,\ \forall s \in S$$

3) Objective 3: Maximising the local incentives

Meaningful local incentives can be defined in two ways. Firstly, by identifying the important factors which cause greatest impact on the location-allocation problem and then secondly, by applying a weighting to each factor, according to importance, and then scoring the factors on each candidate location. The weighted average of these scores can then be defined as the local incentive of each candidate location. The intention then is to maximise the average local incentive of all chosen locations, which would be achieved as follows:

$$\max_{x \in \Omega} TLI = (\Sigma_{\forall w \in W} LI_W Y_W) / (\Sigma_{\forall w \in W} Y_W) + (\Sigma_{\forall d \in D} LI_d Y_d) / (\Sigma_{\forall d \in D} Y_d)$$

4) Objective 4: Minimising the local transportation times

A reduction in transportation time is also an important consideration because not only can it reduce inventory levels, it can also improve customer service levels. The total transport time is therefore the objective function to be minimised, i.e.

$$\text{Min}_{x\in\Omega}\, OTT = \sum_{\forall p\in P}\sum_{\forall w\in W} TT_{pw}X_{pw} + \sum_{\forall w\in W}\sum_{\forall d\in D} TT_{wd}Y_{wd} + \sum_{\forall d\in D}\sum_{\forall c\in C} TT_{dc}Y_{wddc}$$

In summary, the mid-term supply chain planning model can be constructed as a multi-objective-mixed-integer-linear program

4.6.4 Supply Chain Optimisation under Uncertain Demand

The conventional approach for solving a multi-objective optimisation problems is by means of the fuzzy programming approach that can provide a single, yet unprejudiced, final decision and is described accordingly below:

1) Determine the membership function for each fuzzy objective based on the expected upper and lower bounds of the objective values
2) Considering all fuzzy objectives and using the minimum operator, maximising the degree of satisfaction for the worst situation
3) Applying the average operator, maximising the overall satisfaction level with guaranteed minimal fulfilment for all fuzzy objectives

It is noted, in this optimisation approach, that certain parameters are not usually independent and therefore that they cannot be completely separated. Therefore, although an overall solution is not guaranteed, the proposed strategy can provide a definite procedure to reach a single compensatory solution amongst all the participants in a supply chain.

4.6.5 Extraction of the 'Optimisation of a Supply Chain Network under Multi-objective Fuzzy Uncertainty' Technique

An analysis of the Paper, '*Multi-criteria fuzzy optimisation for locating warehouses and distribution centres in a supply chain network*' by Chen et al. (2007) reveals that the basic optimisation under multi-objective fuzzy uncertainty technique employed may be summarised in Table 4-28

Table 4-28 Optimisation under Multi-objective and Fuzzy Uncertainty based on the work of Chen et al. (2007)

1 Problem description of the supply chain network including supply, production, warehousing, distribution and sales

2 Illustrated diagram of supply chain network

3 Discrete demand scenario's with known probabilities modelled as a MILP

4 Specification of all parameters and variables involved in the modelling of the supply chain network

5 Supply Chain modelling (MILP) with demand uncertainty
   1) Specify and define all constraints
      - Network structure constraints
      - Transport constraints
      - Material balance constraints
      - Cost Constraints
   2) Specify and define multiple objectives for optimal planning
      - Minimising total cost
      - Maximising the robustness to various demand scenarios
      - Maximising the local incentives

6 The conventional approaches for solving the multi-objective optimisation problems are usually by means of the fuzzy optimisation approach that can provide a single, yet unprejudiced, final decision as described as follows:
      - Determine the membership function for each fuzzy objective based on the expected upper and lower bounds of the objective values
      - Considering all fuzzy objectives and using the minimum operator, maximising the degree of satisfaction for the worst situation
      - Applying the average operator, maximising the overall satisfaction level with guaranteed minimal fulfilment for all fuzzy objectives

7 Generation of optimum results

4.6.6 Conclusion

This research work by Chen et al. (2007) investigated the multi-objective fuzzy optimisation of a supply chain under uncertainty and was modelled as discrete demand scenarios with given probabilities for different expected outcomes. The supply chain, under uncertainty, optimisation technique utilised in this procedure was extracted and is recorded in Table 4-28.

4.7 Optimisation of a Supply Chain Network under Multi-objective Stochastic Fuzzy Uncertainty

- based on the work of Grossmann and You (2008) -

4.7.1 Introduction

In a paper by Grossmann and You (2008), it was recognised that businesses do not compete as stand-alone entities, but rather as an integral part of supply chain networks whose successes or failures in the market are dependent on end consumer satisfaction. This has resulted in an increased focus on a number of different supply chain related measurable's such as forecasting, inventory levels and supply chain responsiveness. Since responsive supply chains have become a key to competitive success and survival in current times, sophisticated techniques, such as time series, have been developed to improve factors such as forecasting accuracy. Demand uncertainties remain unavoidable on account of ever changing market conditions, and even though dynamic inventory levels may provide some flexibility, the cost thereof may be too exorbitant.

The work of Grossmann and You (2008) considered the design of responsive supply chains with the incorporation of safety stock and inventory levels under demand uncertainty, and with the major goal being the determination of optimal supply chain network configuration in terms of responsiveness and profitability considerations.

In order to properly manage safety stock and inventory levels under demand uncertainty, a probabilistic model for stock-out was followed, mainly because it introduced a quantitative characterisation of responsiveness for supply chain networks. Such probabilistic model could also predict safety stock levels by integrating stock-out probability with demand uncertainty. This is achieved by the formulation and solution of a MINLP (Mixed Integer Non-Linear Program) model.

4.7.2 Quantitative Supply Chain Responsiveness

Since a major goal of this research work was the development of a quantitative definition of supply chain responsiveness incorporating inventory level and demand uncertainty, it was first necessary to define a quantitative measure for supply chain responsiveness.

Since lead time is the time taken for a supply chain network to respond to customer demand, the challenge then is to quantitatively define 'expected lead time'. This is achieved with the integration of supply chain network infrastructure with inventory levels and operating criteria, all under demand uncertainty

4.7.3 Integrated Supply Chain Network Design under Inventory Management and Demand Uncertainty

The Process Supply Chain Network (PSCN), illustrated in Fig 4-10, includes suppliers, plants, distribution centres and customer zones. For the purposes of designing a responsive supply chain network, one operational performance objective would be to minimise the expected lead time of the entire supply chain network, whilst from an economics perspective, while another objective could be to maximise the Net Present Value (NPV) of cash flows over a specified time horizon. Since these two conflicting objectives need to be optimised simultaneously, the corresponding problem will yield an infinite set of trade-off solutions.

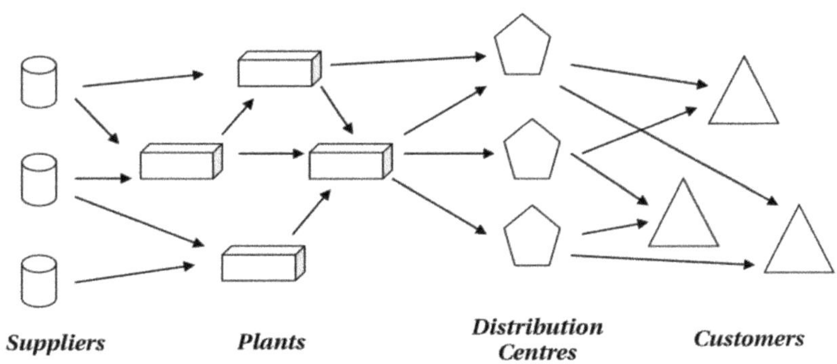

Figure 4-10: Process Supply Chain Network

The model is formulated as a multi-period MINLP problem, which will enable the determination of detailed design decisions, the network infrastructure configuration, production and inventory level operating profiles and various different schedules of the PCSN pertaining to corresponding lead time expectations.

4.7.4 Process Supply Chain Network Model

The model formulated by Grossmann and You (2008) is a multi-period MINLP problem that predicts the network infrastructure, production flows, inventory profiles and schedules of the PSCN with different specifications of the expected lead time. Four types of constraints are included in this model. They are network structure constraints, operational planning constraints, cyclic scheduling constraints and probabilistic constraints. The nomenclature utilised in the model is described in Table 4-28

Table 4-29 Nomenclature for Multi-objective Stochastic Fuzzy Modelling –Grossmann and You (2008)

Index/Set	Physical Meaning
$c \in C$	Customer Zone
$d \in D$	Distribution Centres
$i \in I$	Products
$k \in K$	Transport Capacity Level
$m \in M$	All Objectives
$p \in P$	Plants
$s \in S$	Scenarios
$t \in T$	Periods
$w \in W$	Warehouses
Parameters	Physical Meaning
FCD^{i}	Forecasting customer demand *i* for customer *c*
FTC^{k_*}	k^{th} level fix transport cost, *p to w, w to d, d to c*
LL_*	Local incentive of *w, d*
PPD_s	Probability of product scenario *s*
PQ^*_{ips}	Maximum, minimum manufacturing quantity of product *i*
$Q^{max}{}_{*s}$	Maximum transport quantity of *p* to *w, w* to *d, d* to *c*

$Q^{min}{}_{*s}$	Minimum transport quantity of *p* to *w*, *w* to *d*, *d* to *c*
R_{*nts}	Total resource *n* at *p*
SQ^{+*}	Maximum, minimum capacity of *w*, *d*, + ∈ [max, min}
$TCL^{k}{}_{*}$	*k*th transport capacity level, *p* to *w*, *w* to *d*, *d* to *c*
TT_{*}	Transport time of *p* to *w*, *w* to *d*, *d* to *c*
UEC_{*}	Unit establishing cost of *w*, *d*
$UHC^{i}{}_{*}$	Unit handling cost of product *i* for *w*, *d*
$UPC^{i}{}_{*}$	Unit production cost of product *i* for plant *p*
$UTC^{k}{}_{*}$	*k*th level unit transport cost, *p* to *w*, *w* to *d*, *d* to *c*
$\alpha^{i}{}_{*}$	coefficient relating the capacity of *d* to flow of product *i* handled
$\beta^{i}{}_{*}$	coefficient relating the capacity of *w* to flow of product *i* handled
$\rho^{i}{}_{*nts}$	coefficient for resource *m* used in plant *p* for product *i*
Continuous Variables	Physical Meaning
J_{*}	Objectives
OTT	Total transportation time
$PQ^{i}{}_{*ts}$	Manufacture quantity of *i*
$Q^{i}{}_{*ts}$	Total transport quantity, *p* to *w*, *w* to *d*, *d* to *c*
SQ_{*}	Capacity of *w*, *d*
TCO	Total Cost
TEC	Total establishment cost of all warehouses and distribution centres
THC_{ts}	Total handling cost of scenario *s* in period *t*
TLI, LD	Total local incentive and overall index on distribution centres
TPC_{ts}	Total production cost of scenario *s* in period *t*
TTC_{ts}	Total transportation cost of scenario *s* in period *t*
TTC_{*ts}	Total transportation cost of * for scenario *s* in period *t*
TTQ_{*ts}	*k*th level transport quantity, *p* to *w*, *w* to *d*, *d* to *c*
Binary Var.	Physical Meaning
X_{*}	A link between *p* and *w*, *w* and *d*, *d* and *c* exists
Y_{*}	Warehouse *w* or distribution centre *d* is to be established
$Z^{k}{}_{*ts}$	*k*th transport capacity level, *p* to *w*, *w* to *d*, *d* to *c*
Fuzzy Var.	Physical Meaning
FD	Fuzzy set for final decision
τ_{m}	Fuzzy set for objective *m*

a) Network structure constraints

The determination of the topology of the network infrastructure and the modelling of the inter-connected suite of suppliers, plant sites and distribution centres, where inter-connectedness is represented by the binary variables ($Y^P_{k,l}$, Y^m, $Y^I_{k,Is}$, $Y^N_{k,k'}$ $Y^D_{k,m}$, $Y^S_{m,Id}$) for products (P), distribution centres (m), plants (i), transportation links (N), customers (D), suppliers (S) and manufacturing sites (k). Four sets of network structure constraints are applied to represent the relationships between nodes in the supply chain network.

1) Supplier /Plant-site node

There is a relationship is between suppliers (Is_i) and plant sites (i) (Fig 4-12). A transportation link (N) for raw material j from supplier ls_i to plant site k exists ($Y^I_{k,Is}$), only if at least one plant that consumes raw material j exists in site k ($Y^P_{k,l}$). The relationships discussed above can be expressed by the following logic proposition:

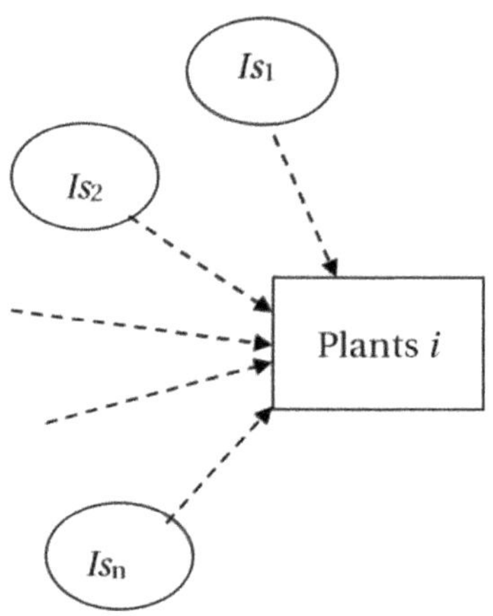

Figure 4-11: Relationship between suppliers and manufacturing sites

$Y^{l}_{k,Is} \rightarrow \vee_{i \in Ij} Y^{P}_{k,I_i}, \forall k \in K, Is \in LS_j, j \in J$

These logic propositions can be further transformed into inequalities:

$Y^{l}_{k,Is} \leq \sum_{I \in Ij} Y^{P}_{k,I}, \forall k \in K, Is \in LS_j, j \in J$ (1)

On the plant side, if a plant that consumes raw material *j* is set up ($Y^{P}_{k,i}$), at least one transportation link from that supplier *ls* to site *k* must be selected ($Y^{l}_{k,Is}$). The logic propositions are:

$Y^{P}_{k,I} \rightarrow \vee_{Is \in LSj} Y^{P}_{k,I_i}, \forall k \in K, i \in I_j, j \in J$

, that can be transformed into inequalities:

$Y^{P}_{k,I} \leq \sum_{I \in LSj} Y^{l}_{k,Is}, \forall k \in K_i, i \in I_j, j \in J$
(2)

2) Input and output / Plant node

There is a second node structure relationship that deals with the input and output flows of a plant (Fig. 4-12). This type of relationship is somewhat more complicated than the one dealing with supplier/manufacturing site relationships because the inter-site transportation link has to be taken into account. If an inter-site transportation link from site k to site k' is installed for chemical j ($Y^N{}_{k,k'}$), it implies that at least one plant i' in site k' is installed that produces chemical j ($_{\mathbf{LSj}} Y^P{}_{k,i'}$), and also at least one plant i in site k is installed that consumes chemical j ($Y^P{}_{k,i'}$)

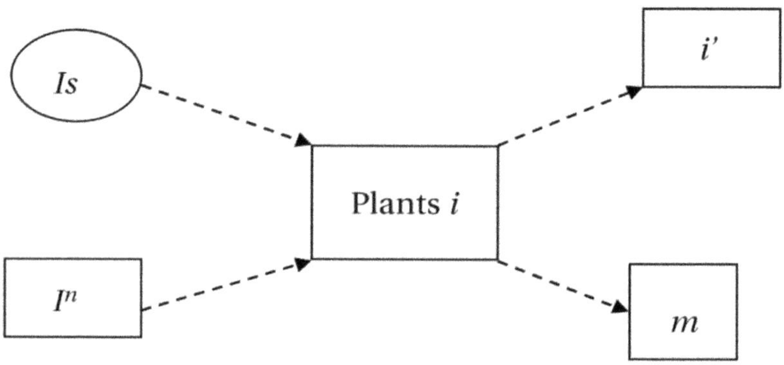

Figure 4-12 Input and output relationship of a plant.

$$Y^N{}_{k',k} \leq \sum_{i \varepsilon Dj} Y^P{}_{k,i}, \ \forall k \varepsilon K, i \varepsilon I_j, j \varepsilon (JP_{k'j} \cap JM_{k,i}) \quad (3)$$

$$Y^N{}_{k',k} \leq \sum_{i \varepsilon Ij} Y^P{}_{k,i}, \ \forall k \varepsilon K, i \varepsilon I_j, j \varepsilon (JP_{k'j} \cap JM_{k,i}) \quad (4)$$

If a plant i in site k is installed ($Y^P_{k,I}$), that consumes chemical j, then site k must be connected to one of the suppliers of chemical j denoted as ls ($Y^l_{k,ls}$), or connected to another site k' that produces chemical j ($Y^N_{k',k}$), or there is another plant i' in site k that produces chemical j ($Y^P_{k,I'}$). The logic propositions can be written as follows:

$$Y^P_{k,I} \rightarrow \vee_{ls \in LSj} Y^l_{k\,ls} \vee_{k' \in ki} Y^N_{k',k} \vee_{i'' \in Dj} Y^P_{k,I}, \forall k \in K, i \in I_j, j \in JM_{k,i}$$

, which can then be transformed into linear inequalities, as follows:

$$Y^P_{k,I} \leq \Sigma_{i \in LSj} Y^l_{k,ls} + \Sigma_{k' \in kj} Y^N_{k',k} + \Sigma_{i \in Dj} Y^P_{k,I'} \forall k \in K, i \in I_j, j \in JM_{k,I} \quad (5)$$

Similarly, if the chemical j is produced by plant i in site k, then at least one other plant i' in the same site is installed that consumes chemical j ($Y^P_{k,I'}$), or there is at least one transportation link to a distribution centre ($Y^D_{k,m}$) or to another site ($Y^N_{k',k}$) that consumes chemical j:

$$Y^P_{k,i} \leq \Sigma_{m \in M} Y^D_{k,m} + \Sigma_{k' \in kj} Y^N_{k',k} + \Sigma_{i \in Ij} Y^P_{k,I'} \forall k \in K, i \in O, j \in JP_{k,I} \quad (6)$$

Constraint (5) is defined for raw materials and intermediate products, and constraint (6) is defined for intermediate and final products.

3) Plant-site / Distribution centre node

The third type of relationship is between plant sites and distribution centres as shown in Fig. 4-13. A transportation link for product *j* from plant site *k* to distribution centre *m* exist ($Y^D_{k,m}$), only if at least one plant that consumes raw material *j* exists in site *k* ($Y^P_{k,I}$). On the plant side, if a plant that consumes raw material *j* is set up, there should be at least one link from the distribution centre *m* to site *k*.

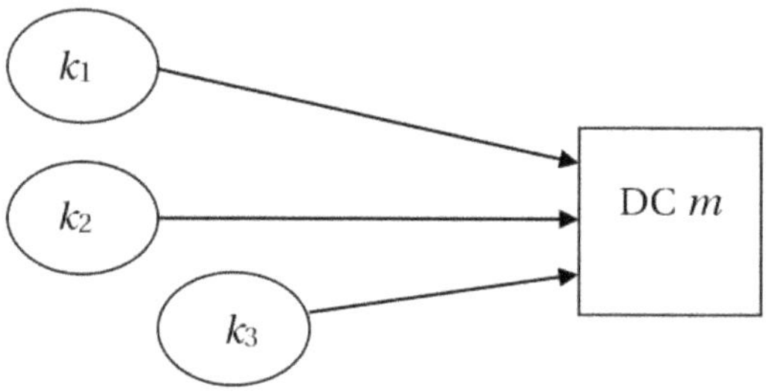

Figure 4-13: Relationship between manufacturing sites, ki and distribution centres,

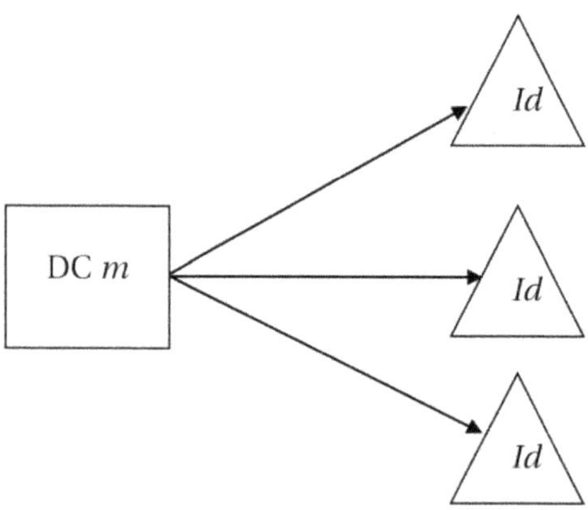

Figure 4-14 Input and output relationship of a distribution centre.

Similarly, transforming the corresponding logic propositions, leads to the following inequalities:

$$Y^{D}{}_{k,m} \leq \sum_{i\,\varepsilon\, Oj} Y^{P}{}_{k,i} \quad \forall k\,\varepsilon\, K,\ m\,\varepsilon\, M,\ j\,\varepsilon\, \mathrm{JP}_{k,i} \tag{7}$$

$$Y^{P}{}_{k,i} \leq \sum_{m\,\varepsilon\, M} Y^{D}{}_{k,m} \quad \forall k\,\varepsilon\, K,\ i\,\varepsilon\, O_j,\ j\,\varepsilon\, \mathrm{JP}_{k,i} \tag{8}$$

4) *Input and output / Distribution centre node*

The last type of network structure relationship is the input and output relationship of a distribution centre as in Fig. 4-14. A transportation link from plant site k to distribution centre m exists, only if the distribution centre m exists (Y^m). A transportation link from distribution centre m to customer ld exists ($Y^{S}{}_{m,Id}$), only if the distribution centre m exists. These binary relationships can be expressed as:

$$Y^{D}{}_{k,m} \leq Y^{m},\ \forall k\,\varepsilon\, K,\ m\,\varepsilon\, M \tag{9}$$

$$Y^{S}{}_{m,Id} \leq Y^{m},\ \forall m\,\varepsilon\, M,\ Id\,\varepsilon\, LD \tag{10}$$

On the other hand, if a distribution centre m is set up, then at least one transportation link from the plant site k to distribution centre m ($Y^{O}{}_{k,m}$) and at least one transportation link from distribution centre m to customer ld ($Y^{S}{}_{m,Id}$) must exist.

$$\mathrm{Y}^{\mathrm{m}} \leq \sum_{k\,\varepsilon\, K} Y^{O}{}_{k,m} \quad \forall m\,\varepsilon\, M \tag{11}$$

$$\mathrm{Y}^{\mathrm{m}} \leq \sum_{k\,\varepsilon\, K} Y^{O}{}_{k,m} \quad \forall m\,\varepsilon\, M \tag{12}$$

b) Operational planning constraints

From an operational perspective, plant capacity, procurement, sales, production, transportation and mass balance relationships are considered together with the constraints of these activities in accordance with the supply chain infrastructure.

1) Production constraints

All the chemical flows $W_{k,I,j,s,t}$ of chemical *j* associated with production scheme *s* in plant *i* of site *k*, other than the main product *j*, are given by the mass balance coefficient $\mu_{i,j,s}$. The following equation relates input or inlet flow of chemical *j* ($W_{k,I,j,s,t}$) with the output of a main product *j* ($W_{k,I,j',s,t}$) of each process:

$$W_{k,i,j,s,t} = \mu_{i,j,s} W_{k,i,j',s,t} \quad \forall k \varepsilon K, i \varepsilon I_j, j \varepsilon JP_{i,s}, s \varepsilon S_i, t \varepsilon T \qquad (13)$$

The production amount should not exceed the design capacity $Q_{k,j}$ defined by the main product *j* for each process:

$$W_{k,i,j,s,t} \leq \rho^1_{i,s} Q_{k,j} Lenp_t \quad \forall k \varepsilon K, i \varepsilon O_j, j \varepsilon JP_{k,i}, s \varepsilon S_i, t \varepsilon T \qquad (14)$$

, where $Lenp_t$ is the length of time period *t* and ρ^1_{is} is the relative production amount of main product *j* of production scheme *s* in plant *i* in turns of capacity.

This formulation is based on the assumption that there are no capacity increments over the entire time horizon. However, multi-period capacity planning events and associated decision variables can readily be considered by appropriately modifying the formulation constraints along with the detailed capacity investment constraints, which are not detailed here. For flexible processes, the maximum production rate $r_{k,i,s}$ of each production scheme *s* in plant *i* of site *k* is proportional to the capacity of the plant i,s is the relative maximum production rate of main product of production scheme *s* in turns of the capacity of plant.

$$r_{k,i,s} = \rho^2_{i,s} Q_{k,j} \quad \forall k \varepsilon K, i \varepsilon O_j, j \varepsilon JP_{k,i}, s \varepsilon S_i, \qquad (15)$$

where $\rho^2_{i,s}$ is the relative maximum production rate of main product of production scheme *s* in terms of the capacity of plant *i*.

2) Mass balance constraints

The mass balance for chemical *j* in manufacturing site *k* at time period *t* is given as follows:

$$\sum_{Is \in LS} P_{k,j,Is,t} + \sum_{k \in K} TR_{k,k',j,,t} + \sum_{i \in Oj} \sum_{s \in S} W_{k,i,j,s,t} =$$
$$= \sum_{m \in M} F_{k,j,m,t} + \sum_{k' \in K} TR_{k,k',j,,t} + \sum_{i \in Ij} \sum_{s \in S} W_{k,p,j,s,t} \quad \forall k \varepsilon K, j \varepsilon J, t \varepsilon T \tag{16}$$

, where $P_{k,j,Is,t}$ is the purchase amount, $TR_{k,k',j,,t}$ is the inter-site shipping amount and $W_{k,i,j,s,t}$ is the production amount. The mass balance for chemical *j* in distribution centre *m* at time period *t* is given as follows:

$$\sum_{k \in K} F_{k,j,m,t} = \sum_{Id \in LD} S_{j,m,Id,t} \quad \forall j \varepsilon J, m \varepsilon M, t \varepsilon T \tag{17}$$

where $F_{k,j,m,t}$ is the shipping amount from production site to distribution centre and $S_{j,m,Id,t}$ is the sales amount.

3) Inventory constraints

The total available amount of chemical *j* for customer *ld* ($QS_{j,Id,t}$) should be equal to the safety stocks ($I_{j,m,Id,t}$) committed to this customer in all distribution centres plus the sale amount.

$$QS_{j,Id,t} = \sum_{m \in M} I_{j,m,Id,t} + \sum_{m \in M} S_{j,m,Id,t}, \quad \forall j \varepsilon J, Id \varepsilon LD, t \varepsilon T \tag{18}$$

The sale amount of chemical *j* to each customer ld at time period *t* should be equal to the associated target demand d^m (usually the target is equal to the mean value of uncertain demand),

$$\sum_{m \in M} S_{j,m,Id,t} = d^m_{j,kl,t}, \quad \forall j \varepsilon J, Id \varepsilon LD, t \varepsilon T \tag{19}$$

The total available amount of chemical *j* will be less than the upper bound of the demand,

$QS_{j,Id,t} \leq d^J_{j,Id,t}, \quad \forall\, j \,\varepsilon\, J,\, Id \,\varepsilon\, LD,\, t \,\varepsilon\, T$
(20)

The working inventories of the plant sites ($WI^{PS}_{k,j,t}$) and the distribution centres ($WI^{DC}_{m,j,t}$) are linearly related to the inlet flows of materials that they handle. This is expressed via the constraints:

$$WI^{PS}_{k,j,t} \geq \alpha_{k,j,t} \textstyle\sum_{Is\, \varepsilon\, LS} P_{k,j,Is,t} \quad \forall\, k \,\varepsilon\, K,\, j \,\varepsilon\, JM_k,\, t \,\varepsilon\, T \qquad (21)$$

$$WI^{DC}_{m,j,t} \geq \beta_{m,j,t} \textstyle\sum_{k\, \varepsilon\, K} F_{k,j,m,t} \quad \forall\, m \,\varepsilon\, M,\, j \,\varepsilon\, JP_k,\, t \,\varepsilon\, T \qquad (22)$$

, where $\alpha_{k,j,t}$ and $\beta_{m,j,t}$ are given coefficients coming from empirical studies (such as the probability of machines broken down or supply limitation).

4) Upper bound constraints

Purchases $P_{k,j,Is,t}$ from supplier *Is* to plant site *k* take place only if the transportation link between them is set up,

$$P_{k,j,Is,t} \leq P^U_{k,j,Is,t} Y^I_{k,Is} \quad \forall\, k, j, Is, t \qquad (23.1)$$

Inter-site transportations $TR_{k,k',j,t}$ from site *k* to site *k'* take place only if the transportation link between them is set up,

$$TR_{k,k',j,t} \leq TR^U_{k,k',j,t}\, Y^N_{k,Is} \quad \forall\, k, k', j, t \qquad (23.2)$$

Sales $S_{j,m,Id,t}$ from distribution centre *m* to customer *ld* take place only if the transportation link between them is selected,

$$S_{j,m,Id,t} \leq S^U_{j,m,Id,t} \quad \forall\, j, m, Id, t \qquad (23.3)$$

Non-zero production flows $W_{k,I,j,s,t}$ are allowed in plant *i* of site *k* only if the plant is installed,

$$W_{k,ij,s,t} \leq W^U_{k,I,j,s,t}\, Y^P_{k,j} \quad \forall\, k, i, j, s, t \qquad (23.4)$$

$$Q_{k,I} \leq Q^{U}_{k,I} Y^{P}_{k,I} \quad \forall\, k, I \tag{23.5}$$

The transportation amount $F_{k,j,m,t}$ from plant site k to distribution centre m takes place only if the transportation link between them is set up:

$$F_{k,j,m,t} \leq F^{U}_{k,j,m,t} Y^{D}_{k,m} \ \forall\, k, j, m, t \tag{23.6}$$

c) Cyclic scheduling constraints

The detailed operations of the multiproduct plants are addressed by the implementation of a cyclic scheduling policy. By this policy, the sequences for producing each product are determined together with cycle time, and then the identical schedule is repeated over each different time period. Optimising of cycle times establishes trade-offs between inventories and product level transitions. Additionally, Important decisions in cyclic scheduling, such as the sequence of production scheduling ($SY_{k,I,s,sl,t}$) and the precedence relationship for changeovers between pairs of production schemes ($Z_{k,I,s,sl,t}$), are determined by the assignment and sequencing of constraints.

1) Assignment constraints

The assignment constraints reflect that only one time slot must be assigned to one production scheme and vice versa. The total number of time slots will be exactly equal to the total number of production schemes:

$$\Sigma_{sl\,\varepsilon\,SL} SY_{k,I,s,sl,t} = 1 \quad \forall k \,\varepsilon\, K, i \,\varepsilon\, I_j, s, t \,\varepsilon\, T \tag{24.1}$$

$$\Sigma_{sl\,\varepsilon\,Si}\, SY_{k,I,s,sl,t} = 1 \quad \forall k \,\varepsilon\, K, i \,\varepsilon\, I_j, s, t \,\varepsilon\, T \tag{24.2}$$

2) Sequence constraints

The sequence constraints reflect the fact that exactly one transition from production scheme s occurs in the beginning of any time slot if, and only if, s was being processed during the previous time slot. On the other hand, exactly one transition to production scheme *s* occurs in the time slot if and only if production scheme *s* is being processed during that same time slot. Transition variables $Z_{k,I,s,s',sl,t}$ can be replaced by continuous variables, ranging between 0 and 1, instead of by binary variables. This significantly reduces the number of discrete variables and improves the computational efficiency.

$$\sum_{s \varepsilon Si} Z_{k,I,s,s',sl,t} = SY_{k,I,s',sl-1,t} \quad \forall k \varepsilon K_i,\ i \varepsilon I_j,\ s' \varepsilon S_i,\ sl \varepsilon SL_i,\ t \varepsilon T \quad (25)$$

$$\sum_{s \varepsilon S'i} Z_{k,I,s,s',sl,t} = SY_{k,I,s',sl,t} \quad \forall k \varepsilon K_i,\ i \varepsilon I_j,\ s' \varepsilon S_i,\ sl \varepsilon SL_i,\ t \varepsilon T \quad (26)$$

$$0 \leq Z_{k,I,s,s',sl,t} \leq 1, \quad \forall k,\ i,\ s,\ s',\ sl,\ t \quad (27)$$

3) Production constraints

The production amount of production scheme *s* in a cycle ($W^s_{k,I,s,t}$) is equal to the processing rate $r_{k,I,s}$ times the processing time $\delta_{k,I,s,s',sl,t}$

$$W^s_{k,I,s,t} = \sum_{sl \varepsilon SLi} r_{k,I,s}\delta_{k,I,s,s',sl,t} \quad \forall k \varepsilon K_i,\ i \varepsilon I_j,\ s \varepsilon S_i,\ t \varepsilon T \quad (28)$$

The amount produced for each product in time period *t* ($N_{k,I,t}$ cycles in the time period) should be no less than the total production predicted from operational planning in this time period:

$$W^s_{k,I,s,t} N_{k,I,t} \geq W_{k,i,j,s,t}, \quad \forall k \varepsilon K_i,\ i \varepsilon O,,\ j \varepsilon JP_{k,i}\ s \varepsilon S_i,\ t \varepsilon T \quad (29)$$

4) Timing constraints

Constraints (30) – (33) are used to restrict timing concerns with cyclic scheduling. The processing time $\delta_{k,I,s,s',sl,t}$ in certain time slots is equal to the summation of the processing times assigned to all the products in those time slots:

$$.\ \delta_{k,i,sl,t} = \sum_{s \varepsilon Si}\delta_{k,I,s,s',sl,t} \quad \forall k \varepsilon K_i,\ i \varepsilon I_j,,\ sl \varepsilon SL_i,\ t \varepsilon T \quad (30)$$

The cycle time $TC_{k,i,t}$ is equal to the summation of all the processing times in each time slot plus the summation of transition times in this cycle,

$$TC_{k,i,t} = \sum_{sl \in SLi} \delta_{k,i,sl,t} + \sum_{s \in Si} \sum_{s' \in Si} \sum_{sl \in SLi} Z_{k,I,s,s',sl,t} \quad (31)$$

$\forall k \in K_i,\ i \in I_j,,\ t \in T$

The total production time should not exceed the duration of each time period $H_{k,i,t}$

$$TC_{k,i,t}\, N_{k,I,t} \leq \quad H_{k,i,t} \qquad \forall k \in K_i,\ i \in I_j,,\ t \in T \quad (32)$$

The production for scheme *s* in time slot *sl* can take place only if the time slot is assigned to the production scheme:

$$\delta_{k,i,sl,t} \leq \delta^{U}_{k,i,sl,t}\, SY_{k,I,s',sl,t} \quad \forall k \in K_i,\ i \in I_j,,\ sl \in SL_i,\ t \in T \quad (33)$$

5) *Cost constraints*

The integration of cyclic scheduling with the strategic planning required the consideration of inventory and transition costs from cyclic as part of operating costs. The total operating cost of a certain multi-product plant producing product *a,* in a particular time period, is represented by constraint (34). The first term on the right hand side of the equation represents the total transition cost corresponding to a certain time period. The second term of the constraint is the inventory cost for all chemicals involved in the production process. The change of inventory level over a certain time period is given in Fig. 10, which shows that each manufacturing site may have more than one production plant, and that inventory levels for materials at multi-product plants are also taken into account. Since it is further assumed that material balances are expressed linearly in terms of the product production, the cumulative inventory levels for raw materials are therefore also linearly related to the cumulative inventory levels of main product in each production scheme. Consequently, the coefficients of the linear relationships are taken to be equal to the mass balance coefficients. Therefore, the inventory and transition costs of multiproduct processes are given by:

$$\begin{aligned} COSTS^{S}_{k,i,t} = & \sum_{s \in Si} \sum_{s' \in Si} \sum_{sl \in SLi} CTR_{i,s,s'}\, Z_{k,I,s,s',sl,t}\, N_{k,I,t} + \\ & \sum_{s \in Si} \sum_{s' \in Si} \sum_{sl \in SLi} (\mu_{i,j,s}\, \varepsilon_{k,j,t})\,(r_{k,I,s}\, H_{k,i,t}) - \\ & W^{s}_{k,I,s,t} \quad N_{k,I,t} \delta_{k,i,sl,t}/2 \qquad \forall k \in K_i,\ i \in I_j,,\ sl \in SL_i,\ t \in T \end{aligned}$$

(34)

This constraint is non-linear and non-convex, with bilinear and tri-linear terms. If all the processes in the production network are dedicated, cyclic scheduling need not be taken into account, and this constraint can therefore be discarded.

6) Upper bound constraints

As a multi-site problem, it must be ensured that if a plant *i* is not installed in site *k*, then there are no production cycles. To model this, an upper bound constraint (35) is introduced the for this number of cycles $N_{k,I,t}$ in each time period for each multiproduct plant in each manufacturing site:

$$N_{k,I,t} \leq N^{U}_{k,I,t} Y^{P}_{k,I} \qquad \forall k \varepsilon K_i, i \varepsilon I_j, s \varepsilon S_i, t \varepsilon T \qquad (35)$$

Also the assignment constraints are reformulated to account for the fact that all the scheduling activities can take place only if the plant is installed:

$$\sum_{sl \, \varepsilon \, SLi} SY_{k,I,s',sl,t} = Y^{P}_{k,I} \qquad \forall k \varepsilon K_i, i \varepsilon I_j, s \varepsilon S_i, t \varepsilon T \qquad (36.1)$$

$$\sum_{sl \, \varepsilon \, Si} SY_{k,I,s',sl,t} = Y^{P}_{k,I} \qquad \forall k \varepsilon K_i, i \varepsilon I_j, sl \varepsilon SL_i, t \varepsilon T \qquad (36.2)$$

d) Probabilistic constraints for demand uncertainty

Since a key component of decision-making under uncertainty is the representation of the stochastic parameters, there are two distinct ways of representing such uncertainties in modelling endeavours. There is the scenario-based approach that attempts to accommodate uncertainties by representing them in terms of discrete realisations of the stochastic parameters where each complete realisation of all uncertain parameters gives rise to a scenario. In this way all the possible future outcomes are taken into account through the use of scenarios. This approach provides a straightforward way of formulating the problem, but has a major drawback in that the problem size increases exponentially as the number of scenarios increases. This is particularly true when using continuous multivariate probability distribution with Gaussian quadrature integration schemes.

Alternatively, Monte Carlo sampling could be used, although it requires a fairly large number of sample points to achieve a desired level of accuracy. These difficulties may be overcome by the integration of continuous probability distribution functions as random parameters. Such an approach can lead to manageably sized deterministic equivalent representations of the probabilistic model, but at the expense of introducing a certain number of non-linear relationships into the model through multivariate integration over the continuous probability space. However, in this work, such an approach is used for describing the demand uncertainty and the probabilistic description of demand uncertainty therefore results in some operational planning variables (amount of sale, shortfall and salvage) being stochastic. Alternatively, the traditional way of dealing with this is by allowing "corrective actions" to be introduced by adding recourse actions into the model. Stochastic programming problems with recourse are usually complicated in the nature and difficulties to achieve solution. Based on the special characteristics of demand uncertainty, a novel approach to minimise the effect of unexpected demand is by holding a certain amount of safety stock before the realisation of demand is proposed. The quantity of safety stock required is estimated by applying chance constraints to a demand probability distribution. The need for recourse is removed by taking the proactive action with safety stock, and therefore the stochastic attributes of the problem are translated into a deterministic optimisation scenario at the expense of introducing non-linear terms into the model. In this work, a similar approach is used in that instead of specifying the service level (or over-stocked probability), the stock-out probability is treated as a variable, and then integrated with supply chain responsiveness. The detailed model formulations are given below.

1) Stock-out probability

If a particular product demand realization $d_{j,Id,t}$ is higher than its total amount available, $QS_{j,Id,t}$, then an under-stocked situation arises, i.e. a stock-out situation will occur. If a particular product demand realization $d_{j,Id,t}$ is less than its total available amount $QS_{j,Id,t}$,an over-stocked situation occurs. The probability of an over-stock situation is defined as β-service level in the manufacturing literature. The probability of stock-out plus the service level should be always equal to 1. Thus, the stock-out probability can therefore be expressed as;

$$\text{Prob}_{j,Id,t} = \Pr(QS_{j,Id,t} < d_{j,Id,t})$$

$$= 1 - \Pr(QS_{j,Id,t} \geq d_{j,Id,t}), \quad \forall j \varepsilon J, Id \varepsilon LD, t \varepsilon T \qquad (37)$$

The nature of this constraint is very similar to that of a chance constraint, which suggests that equation (37) can be transformed into a deterministic equivalent. It is worth considering that this general definition for estimating safety stock level can be applied for all types of demand distributions. In real life situations, however, demand is often assumed to be normally distributed, especially when there are sufficient samples, or triangularly distributed ones when limited samples are involved. In the following sections, the scenarios when the demand follows a normal distribution and a triangular distribution are considered.

2) Normal distribution

As a consequence of the central limit theorem, product demands are often modelled as random parameters, which are normally distributed. For normally distributed demand, the mean ($\mu_{j,Id,t}$) and the standard deviation ($\sigma_{j,Id,t}$) of the demand are usually provided. To facilitate the calculation of stock-out probability $\text{Prob}_{j,Id,t}$, standardisation of the variables is initially performed. Normal random variables can be recast into the standardized normal form, with a mean of zero and a variance of 1, by subtracting their mean and dividing by their standard deviation. For the deterministic variables $QS_{j,Id,t}$ the standardized normal variables $K_{j,Id,t}$ are given as:

$$K_{j,Id,t} = (QS_{j,Id,t} - \mu_{j,Id,t}) / \sigma_{j,Id,t} \qquad (38)$$

Substituting (38) into the general definition of stock-out probability defined in (37),we have the stock-out probability of normally distributed demand for product *j* customer *ld* at time period *t* is:

$$\text{Prob}_{j,ld,t} = 1 - \phi(K_{j,ld,t}), \qquad \forall j \in J, ld \in LD, t \in T$$

,where $\phi(K_{j,ld,t})$ denotes the cumulative probability function of standard normal random variable in the form of:

$$\phi(x) = 1/\sqrt{2\pi} \int^{\infty}_{-\infty} \exp(x^2/2).dx$$

The stock-out probability for the entire planning horizon is considered as the worst case for all the individual stock-out probabilities of all the time periods. Hence, we have that the stock-out probability ($\text{Prob}_{j,ld,t}$) for product *j* and customer *ld* is:

$$\text{Prob}_{j,ld} \geq \text{Prob}_{j,ld,t} = 1 - \phi(K_{j,ld,t}), \quad \forall j \in J, ld \in LD, t \in T \qquad (39)$$

3) Triangular distribution

For triangular distribution, the demand's lower bound d^L, upper bound d^U, and the most likely value d^M are given. On account of the non-differentiable characteristics of a triangular distribution, the stock-out probability has different representations when the total available amount *QS* is more or less equal to the most likely demand d^M Their relationship can be represented by the following disjunction (for simplicity, the subscripts *j, ld, t* are omitted)

$$\begin{Bmatrix} d^L \leq QS < d^U \\ \text{Prob} \geq 1 - \dfrac{(QS - d^L)^2}{(d^U - d^L)(d^M - d^L)} \end{Bmatrix} \vee \begin{Bmatrix} d^L \leq QS < d^U \\ \text{Prob} \geq \dfrac{(d^U - QS)^2}{(d^U - d^L)(d^M - d^L)} \end{Bmatrix} \qquad (40)$$

Using the convex hull reformulation, the disjunction can be transformed into MINLP constraints:

$$QS_{j,kl,t} = QS^{1}_{j,kl,t} - QS^{2}_{j,kl,t} \tag{40.1}$$

$$d^{L}_{j,ld,t}\, Y^{d}_{j,kl,t} \leq QS^{1}_{j,kl,t} \leq d^{M}_{j,ld,t}\, Y^{d}_{j,kl,t} \tag{40.2}$$

$$d^{M}_{j,ld,t}(1 - Y^{d}_{j,ld,t}) \leq QS^{2}_{j,ld,t} \leq d^{U}_{j,ld,t}(1 - Y^{d}_{j,kl,t}) \tag{40.3}$$

$$\mathrm{Prob}_{j,ld,t} = \mathrm{Prob}^{1}_{j,ld,t} + \mathrm{Prob}^{2}_{j,ld,t} \tag{40.4}$$

$$\mathrm{Prob}^{1}_{j,ld,t} \leq Y^{d}_{j,ld,t} \tag{40.5}$$

$$\mathrm{Prob}^{2}_{j,ld,t} < 1 - Y^{d}_{j,ld,t} \tag{40.6}$$

$$\mathrm{Prob}^{1}_{j,ld,t} \geq Y^{d}_{j,kl,t} - \left\{ \frac{(QS^{1}_{j,kl,t} - d^{L}_{j,ld,t})}{(d^{U}_{j,ld,t} - d^{L}_{j,ld,t})(d^{M}_{j,ld,t} - d^{L}_{j,ld,t})} \right\}$$

$$\mathrm{Prob}^{2}_{j,ld,t} \geq \left\{ \frac{[(1 - Y^{d}_{j,kl,t})\, d^{U}_{j,ld,t} - QS^{2}_{j,kl,t}]^{2}}{(d^{U}_{j,ld,t} - d^{L}_{j,ld,t})(d^{U}_{j,ld,t} - d^{M}_{j,ld,t})} \right\} \tag{40.7}$$

, where $Y^{d}_{j,kl,t}$ is a binary variable that's equal to 1 if $QS_{j,ld,t}$ is less than $d^{M}_{j,ld,t}$.

4) *Lead time definition*

As discussed in Section 3, the expected lead time is equal to the delivery lead time for each path, which is added to the production lead time, multiplied by the stock-out probability. The delivery lead time and the production lead time are in turn equal to the summation of all the production delays and transportation delays incurred. The expected lead time of the whole supply chain network is equal to the maximum expected lead time of each path. In a supply chain design problem when a plant or a transportation link is not selected, the associated delay is taken to be 0. Binary variables are used to model the lead time TP with the following inequalities:

$$TP \geq Prob_{j,ld}\, Y^{l}_{k,ls} + \Sigma^{n}_{x=1} Prob_{j,ld} Y^{P}_{kx,ix} \theta_{kx,ix} + \Sigma^{n-1}_{x=1}$$

$$Prob_{j,ld} Y^{N}_{kx,kx+1} + Prob_{j,ld} Y^{O}_{kn,m+1} \lambda^{O}_{n,m} + Y^{S}_{m,ld+1} \lambda^{S}_{m,ld} \qquad (41)$$

where $Prob_{j,ld}$ is the stock-out probability, all the *Y* are binary variables for design decisions, λ denotes transportation delays and θ represents production delay. The superscript $(*)^{l}$ denotes the transport link from supplier to plant site, the superscript $(*)^{P}$ denotes the plant, $(*)^{N}$ is for inter-site transport links, $(*)^{O}$ represents the transport link from plant site to distribution centre, and $(*)^{S}$ is for the transport link from distribution centre to customer.

The set $Path_{ls,k,m,ld}$ includes all the possible paths in a given potential PSCN network. All the elements in the set $Path_{ls,k,m,ld}$ are in the form of $(ls, k_1, k_2 \ldots k_n, m, ld)$, where supplier *ls* is the start of the path, $k_1, k_2 \ldots k_n$, are the manufacturing sites and *m* is the distribution centre that the associated stream go through, and customer *ld* is at the end of the path. In Eq. (41) the transportation delay of each route and the production delay of each single product plant are given parameters. The production time delay for a multiproduct plant is not so obvious. Before introducing the definitions, a multiproduct plant that produces three chemicals *A*, *B* and *C* is considered. Assuming there is a demand change of chemical *A*, a worst case occurs when the production of *A* has been completed, and there is no additional inventory of *A* available, besides that already committed to previous demand. Two different operating policies may be implemented to deal with this situation. In the first case, if the demand of chemical *A* has a large positive change, the production of chemical *C* plus the other alternative, Chemical *B*, could be interrupted to produce chemical *A*. In this case, the production delay is equal to the residence time of chemical *A* plus the transition time from producing product *C* to producing product *A*. Therefore, the production delay $\theta_{k,i}$ for multiproduct plant *i* in site *k* given by:

$$\theta_{k,i} \geq \theta^{R}_{k,i} + \tau_{i,s,s'} \quad \forall k \in K_i,\ i \in I_j,\ s,\ s' \in S \tag{42.a}$$

If the demand change of chemical *A* is not very significant, a temporary delay will be incurred until the normal production can be resumed again, thus enabling the adjustment of production to satisfy the change in demand. The amended production time taken is given by the processing time of chemical *B* and *C*, plus residence time of *A*. In this way, the time delay for each product, on the multiproduct plant, is given by cycle time plus residence time less the processing time (Fig. 4-14). Similarly, the production delay for a multiproduct plant is equal to the maximum time delay for each product:

$$\theta_{k,i} \geq TC_{\mathbf{k,i,t}} + \theta^{R}_{\mathbf{i,s}} - \sum_{\mathbf{sl} \in \mathbf{SLi}} \delta_{\mathbf{k,i,s,sl,t}} \quad \forall k \in K_i,\ i \in I_j \tag{42.b}$$

In this definition cycle times of each plant are taken into account as part of the delay due to production, so that production details have been integrated into the quantitative definition of responsiveness.

The terms $Prob_{\mathbf{j,Id}}Y$ (stock-out probability times binary variable omitting the subscripts for simplicity) in the lead time definition can be linearized. A continuous variable PY is used to replace the $Prob_{\mathbf{j,Id}}Y$ term in the lead time constraint:

$$Prob_{\mathbf{j,Id}}Y = PY, \quad \forall j,\ Id \tag{43.1}$$

$$TP \geq PY^{\mathbf{I}}_{\mathbf{k,j,Is,Id}}\lambda^{\mathbf{I}}_{\mathbf{k,Is}} + \sum^{\mathbf{n}}_{\mathbf{x=1}} PY^{\mathbf{P}}_{\mathbf{kx,ix,j,Id}}\theta_{\mathbf{kx,ix}} + \sum^{\mathbf{n-1}}_{\mathbf{x=1}} PY^{\mathbf{N}}_{\mathbf{kx,kx+1,j,Id}}\lambda^{\mathbf{N}}_{\mathbf{kx,I}} + PY^{\mathbf{O}}_{\mathbf{kn,j,m,kl}}\lambda^{\mathbf{N}}_{\mathbf{kx,I}} + Y^{\mathbf{S}}_{\mathbf{m,kl}}\lambda^{\mathbf{S}}_{\mathbf{m,Id}},\ \forall (I_s,\ k_1,\ k_2 \ldots k_n,\ m,\ Id) \in Path_{Is,k,m,Id} \tag{43}$$

Eq. (43.1) is equivalent to the following disjunction:

$$\begin{Bmatrix} Y \\ PY = Prob_{j,Id} \\ \end{Bmatrix} \vee \begin{Bmatrix} -Y \\ PY = 0\,,\ \forall j,\ Id \\ \end{Bmatrix} \quad (44.1)$$

Applying the convex hull reformulation to the above disjunctive constraint leads to:

$$PY + PY2 = Prob_{j,Id},\quad \forall j,\ Id \quad (44.2)$$

$$PY \leq Y,\quad \forall j,\ Id \quad (44.3)$$

$$PY2 \leq 1 - Y,\quad \forall j,\ Id \quad (44.4)$$

,where PY2 is a new continuous variable introduced as a slack variable. The constraints (44) are applied for all the terms with superscript $(*)^{I}$, $(*)^{P}$, $(*)^{N}$, $(*)^{O}$, $(*)^{S}$ in the expected lead time definition.

5) Nonnegative constraints

All continuous variables must be nonnegative and the binary variables should be integer:

$$Q_{k,I},\ W_{k,i,j,t},\ P_{k,j,Is,t},\ TR_{k,k',j,x},\ F_{k,j,m,t},\ S_{j,m,Id,x},\ I_{j,m,Id,x},\ QS_{j,Id,t},\ TP,\ \theta \geq 0 \quad (45.1)$$

$$Z_{k,I,s,s',sl,t},\ W^{S}_{k,I,s,t},\ r_{k,i,s},\ \delta_{k,i,sl,t},\ TC_{k,i,r},\ Te_{k,i,sl,t},\ Ts_{k,i,sl,t},\ N_{k,i,t},\ \theta_{k,i},$$

$$COST^{S}_{k,i,t} \geq 0 \quad (45.2)$$

$$Y^{I}_{k,Is},\ Y^{P}_{k,I},\ Y^{D}_{k,m},\ Y^{m},\ Y^{S}_{m,Id},\ SY_{k,i,s,sl,r}\ \varepsilon\ (0,1) \quad (45.3)$$

6) *Net present value*

The NPV of the supply chain network is given by the following equations:

$$NPV = Income - C_{purch} - C_{oper} - C_{tranp} - C_{invest} - C_{inventory}$$

$$Income = \sum_j \sum_{kl} \sum_t \phi_{j,Id,t} Sa_{j,Id,t}$$

$$C_{purch} = \sum_k \sum_j \sum_{Is} \sum_t \phi_{j,Id,t} P_{k,j,Is,t}$$

$$C_{oper} = \sum_k \sum_j \sum_s \sum_j \sum_t \sigma_{i,s,t} W_{k,i,j,s,t} + \sum_k \sum_i \sum_t COST^S_{k,i,t}$$

$$C_{tranp} = \sum_k \sum_j \sum_{Is} \sum_t w^I_{k,j,Is,t} P_{k,j,Is,t} + \sum_k \sum_{k'} \sum_j \sum_t w^N_{k,k',j,t} TR_{k,k',j,t} + \sum_k \sum_j \sum_m \sum_t w^O_{k,j,m,t} F_{k,j,m,t} + \sum_j \sum_m \sum_{Id} \sum_t w^S_{j,m,Id,t} S_{j,m,Id,t} \quad (46)$$

$$C_{invest} = \sum_k \sum_i w^P_{k,i} Q_{k,i} + \sum_k \sum_i \gamma^P_{k,i} Y^P_{k,i} + \sum_k \sum_{Is} \gamma^I_{k,Is} Y^P_{k,i} + \sum_k \sum_m \gamma^O_{k,m} Y^O_{k,m} + \sum_k \sum_m \gamma^N_{k,k'} Y^N_{k,k'} + \sum_m \gamma^m Y^m + \sum_k \sum_i \gamma^S_{k,Id} Y^S_{k,Id}$$

$$C_{inventory} = \sum_j \sum_m \sum_{Id} \sum_t \varepsilon_{j,m,t} I_{j,m,Id,t} + \sum_k \sum_j \sum_t \varepsilon_{k,j,t} (WI^{PS}_{k,j,t} + WI^{DC}_{m,j,t}$$

All the parameters in the above formulation are discounted at a specified interest rate and include the effect of taxes and interest rate on the net present value.

4.7.5 Solution procedure

a) Solution procedure for multi-objective optimisation

In order to achieve Pareto-optimal curve for this bi-criterion optimisation problem, one of the objectives is defined as an inequality with a fixed value for the bound treated as a parameter. There are two major approaches to solve this problem with regard to this parameter. One such approach is to simply solve it for a specified number of points to obtain an approximation of the Pareto-optimal curve. The other approach is to solve the problem as a parametric programming problem, which will yield the precise solution for the Pareto-optimal curve. Whilst the latter approach provides for a rigorous solution, the former approach is simpler to implement, especially for non-linear models. For this reason, the latter approach was selected and the associated procedure involves the following three steps:

1) The first step is to minimise the expected lead time TP in order to obtain the shortest expected lead time TPS, which in turn yields the lowest Pareto-optimal NPV.

2) The second step is to maximize NPV that in turn yields the longest Pareto optimal expected lead time TPL. In this case the objective function is set as, NPV – $\varepsilon \cdot$ TP (47), where ε is a very small value (e.g. it is on the order of 0.001).

3) The third step is to fix the expected lead time TP to discrete values between TPS and TPL, and optimize the model by maximizing NPV at each selected point. In this way an approximation to the Pareto-optimal curve together with the optimal configurations of PSCN for different values of lead time, is obtained

b) Shortest optimal expected lead time

In the solution procedure for this bi-criterion optimisation problem, described above, one of the important steps is the determination of the shortest optimal expected lead time. Rather than minimising the expected lead time by solving the entire problem directly, the following solution approach is applied to improve computational efficiency. The expected lead time of each path of chemical flow as defined in Section 3.2 is given by $E(L) = \text{Prob} \cdot L_P + L_D$, which equals the delivery lead time (LD) plus stock-out probability (Prob) times production lead time (LP). From the above equation, it can be seen that as the stock-out probability decreases, the expected lead time will decrease. In the step for determining the shortest optimal expected lead time, the economic objective is not accounted for and therefore, if there are sufficiently high safety stock levels in all the distribution centres to hedge the uncertain demands, the stock-out probability will be 0.

In such a case, the expected lead time of the supply chain network will be reduced to the maximum delivery lead time of each path of chemical flow. The delivery lead time is equal to the transportation delay from a distribution centre to a customer. Since the type and number of distribution centres to be installed, and also the extent to which transportation links may be set up, may be selected, a subset may be chosen of all possible transportation links between the customers and distribution centres. In such a way, each customer has at least one transport link connected, and the maximum transportation delay of all these transport links is minimized. In summary, the minimum expected lead time can be calculated by:

$$TP_S = \max_{ID \in LD}(\min_{m \in M} \lambda^S_{m,Id}) \qquad (48)$$

, where $\lambda^S_{m,Id}$ is the shipping time from distribution centre *m* to customer *ld*. Eq. (48) defines the shortest expected lead time, which means that for each customer *ld*, a distribution centre is selected that has minimum shipping time to the customer, and a transportation link is set up between them. In such a way, the number of transport links that are set up is equal to the number of the customers (number of elements in set *ld*) that exist. Since the expected lead time of the supply chain network is equal to the longest expected lead time for each path, the minimum expected lead time of the supply chain network will be equal to the maximum shipping time for those transportation links that have been established. The optimal solution may be obtained by using the proposed method instead of solving the large-scale non-convex MINLP problem, and with much less computational effort. However, this method will only work for the shortest expected lead time scenario. Other points in the Pareto curve may be obtained by solving a series of large-scale non-convex MINLP problems. To reduce the computational time involved, a heuristic hierarchical solution approached is proposed.

c) Heuristic hierarchical approach

The solution of each problem for the points in the Pareto curve can be very computationally demanding, on account of the fact that a large number of discrete decisions and highly non-linear non-convex terms are involved. Therefore a heuristic hierarchical solution approach is proposed that is able to handle the combinatorial and non-convex nature of the responsive supply chain design problem and is also able to reduce the computational effort required. The basic idea exploits the fact that the operational cost arising from scheduling (inventory costs and transition costs) constitutes only a small part of the total NPV and therefore that a two-stage decision approach may be followed. The supply chain network structure is at first determined and then strategic planning decisions (production and transportation amounts), neglecting changeovers and transitions for multiproduct plants, are taken into account. In the second stage the detailed production scheduling, after the strategic design and planning decisions are made, is considered. The proposed technique employs a simplified model (i.e. first stage) in order to determine the strategic infrastructural decisions, which are then fed into the detailed model (i.e. second stage) in order to determine the operational and scheduling decisions. It is important to note that this solution approach can only be applied once the production delay constraints are defined (42.a), which assumes, of course, that the demand is undergoing large changes and that the production delay is irrelevant to the cycle time (please refer to Section 5.5 for details). Clearly this heuristic approach may not yield optimal or near-optimal solutions if the model includes production delay constraints (42.b). The reason is given in Section 6.3.1.

1) Simplified model

The simplified model formulation is an approximation of the detailed model formulation. The main advantage of the simplified model is that it does not include the large amount of non-convex terms and binary variables required for cyclic scheduling. In this way, the problem can be solved more efficiently and still capture the main trends in the supply chain design and planning. Furthermore, this model provides a valid upper bound by overestimating the objective function of the original problem. Also since transitions and changeovers for multiproduct plants are not taken into account in the simplified model, all the constraints for cyclic scheduling are not included the model. The objective functions are the same for each step as in the previous solution procedure for the multi-objective optimisation problem. In summary, the simplified model includes the following constraints: (1)–(23), (37)–(40) (depends on the associated demand probability distribution), (42.a), (44)–(45). The objective functions are (43) and (46).

In the simplified model, the production delay is defined with constraint (42.a), which regards production delay as being equal to the maximum residence time plus transition time of all the products. Consequently, the expected lead time of a supply chain will depend on the design and planning decisions only, and is unrelated to production scheduling. Therefore, the expected lead time, which is obtained from the simplified model, is exactly the same as that obtained by solving the original model. As a result of this only the second step of the ε-constraint method, which is to calculate the longest optimal expected lead time of the simplified model, needs to be solved. On the other hand, the simplified model might not provide a good approximation of the original model if such model includes constraint, (42.b), to define the production delays, i.e. cycle times and processing times, instead of constraint, (42.a), which defines demand-related delays. However, cycle times and processing times, which are decisions for cyclic scheduling, had been neglected in the simplified model and therefore missing production schedule information may lead to inaccuracies when optimising the expected lead time. Therefore a heuristic scheduling approach may not yield near-optimal solutions if the model includes constraint (42.b) for the production delays.

2) Solving detailed model

In the detailed model, the design variables (binary variables for design decisions) and planning variables (production and transportation amounts, inventory levels) are determined from the simplified model. The detailed model is then solved with the simplified variable set in order to determine the optimal levels for the scheduling decision variables. However, at this point, the detailed model becomes a large-scale non-convex MINLP model and therefore to obtain a 'good' starting point with such near-optimal solution set, a heuristic sub-problem and a bounded sub-problem, are formulated. The heuristic sub-problem is used for the production sequence of multi-product plants and is based on the fact that transition costs are usually proportional to the transition times. Consequently, a production sequence that can minimise the total transition time in a production cycle is often the optimal production sequence. Therefore, the heuristic sub-problem is to minimize the total transition times in a production cycle (constraint (31)), subject to the constraints (24)–(27). In the bounded sub-problem, all the bilinear terms in the detailed model are replaced by convex envelopes. For example, the equation $W^{S}{}_{k,I,s,t} N_{k,i,t} = W_{k,i,j,s,t}$ (29) was replaced by the following constraints:

$$W^{S}{}_{k,i,s,t}{}^{LO} N_{k,i,t} + W^{S}{}_{k,i,s,t} N^{LO}{}_{k,i,t} - W^{S}{}_{k,i,s,t}{}^{LO} N^{LO}{}_{k,i,t} \leq W_{k,i,j,s,t} \quad (49.1)$$

$$W^{S}{}_{k,i,s,t}{}^{UP} N_{k,i,t} + W^{S}{}_{k,i,s,t} N^{UP}{}_{k,i,t} - W^{S}{}_{k,i,s,t}{}^{UP} N^{UP}{}_{k,i,t} \leq W_{k,i,j,s,t} \quad (49.2)$$

$$W^{S}{}_{k,i,s,t}{}^{LO} N_{k,i,t} + W^{S}{}_{k,i,s,t} N^{UP}{}_{k,i,t} - W^{S}{}_{k,i,s,t}{}^{LO} N^{UP}{}_{k,i,t} \leq W_{k,i,j,s,t} \quad (49.3)$$

$$W^{S}{}_{k,i,s,t}{}^{UP} N_{k,i,t} + W^{S}{}_{k,i,s,t} N^{LO}{}_{k,i,t} - W^{S}{}_{k,i,s,t}{}^{UP} N^{LO}{}_{k,i,t} \leq W_{k,i,j,s,t} \quad (49.4)$$

,where LO and UP represent the lower and upper bounds of the variables respectively. The solutions of the two sub-problems are then used as the starting point to solve the detailed model. By doing this, the possibility of obtaining a near-optimal solution is increased.

Applied Algorithm

In summary, the hierarchical algorithm consisted of the following steps:

Step 1: Solve the simplified non-convex MINLP model (objective functions (43) and (46), subject to constraints (44)–(45)) by neglecting transitions and changeovers of multiproduct plants and then fix the design and planning decisions.

Step 2: For each multiproduct plant, solve the MILP heuristic Sub-problem (objective function is (31), constraints are (24)–(27)) to minimize the total transition time in a production cycle to obtain the initial values of the scheduling variables.

Step 3: Solve the bounded model that uses convex envelopes to replace the bilinear constraints (29), (32) and (34).

Step 4: Use the solutions from the sub-problems in Step 2 and 3, as the starting points in solving the detailed MINLP model in the reduced space, to determine the scheduling and operational decisions.

4.7.6 Extraction of the 'Supply Chain Network under Multi-*objective* Stochastic Fuzzy Uncertainty' Methodology

An analysis of the Paper, '*Design of responsive supply chains under demand uncertainty* by Grossmann and You (2008) reveals that the optimisation under multi-objective fuzzy uncertainty technique applied may be summarised in Table 4-30.

Table 4-30 Optimisation of a supply chain under a Multi-objective and Fuzzy Uncertainty Technique based on the work of Grossmann and You (2008)

1 Problem description of the supply chain network including supply, production, warehousing, distribution and sales

2 Illustrated diagram of supply chain network

3 Specify the chief requirements of the supply chain network

4 Specification of all parameters and variables involved in the modelling of the supply chain network

5 Supply Chain modelling for a batch processing operation with demand uncertainty
   a) Specify and define all deterministic constraints
      - Network structure constraints
      - Operational planning constraints
      - Cyclic scheduling constraints
      - Cost Constraint
   b) Specify all probabilistic constraints related to demand uncertainty
      - Stock-out probability
      - Demand Uncertainty modelled by either Normal Distribution or Triangular Distribution
      - Lead Time Definition (probabilistic)
   c) Specify all process limitations
   d) Specify and define multiple objectives for optimal planning
      - Minimising lead time
      - Maximising Net Present Value

6 Solution Procedure for batch processing operation (NB: this step wouldn't be necessary for a continuous process operation)
      - Solve the MINLP (Mixed Integer Non-Linear Linear Programming) model for the entire operation by neglecting any transitions or change-overs between constituent plants
      - For each plant, solve the MILP sub-problem to minimise the total transition time in a production cycle to obtain initial values for scheduling variables
      - Solve the bounded model that uses convex envelopes to replace the bi-linear constraints
      - Use the solutions from steps b) and c) as the starting points in solving the detailed MINLP model in the reduced space to determine the optimal scheduling and operational solution

7 Record the optimal solution

4.7.7 Conclusion

In this paper, a quantitative approach for designing responsive supply chains under demand uncertainty was presented. The expected lead time was proposed as a measure of process supply chain responsiveness, and defined quantitatively with integration of supply chain network structure and inventory level. A multi-period mixed integer non-linear programming (MINLP) model was developed for the bi-criterion optimisation of economics and responsiveness, while considering customer demand uncertainty. The model integrated the long-term strategic decisions (e.g. installation of plants, selection of suppliers, manufacturing sites, distribution centres and transportation links) with the short-term operational decisions (e.g. product transitions and changeovers) for the multi-site multi-echelon process supply chain network. The model also included a novel approach for predicting the safety stock levels with consideration of responsiveness, demand uncertainty and economic objectives. A bi-criterion optimisation approach was implemented to obtain the trade-off curve between responsiveness and economics using the ε-constraint method. A hierarchical algorithm was further presented for the solution of the resulting large-scale MINLP problem based on decoupling of the decision-making levels (strategic and operational). Although in this work only demand uncertainty is taken into account, the model can be extended to consider supply uncertainty by introducing safety stocks of the raw materials. A similar probabilistic approach can be used to determine safety stocks levels of different raw materials, but a resulting difficulty is how to quantify the stock-out probability given the presence of both supply and demand uncertainty. Finally the optimisation under uncertainty technique was extracted and is summarised in Table 4-12.

The supply chain, under uncertainty, optimisation technique utilised in this procedure was extracted and is recorded in Table 4-30.

4.8 A Global (Pareto-) Optimisation Strategy for Chemical Supply Chains under uncertainty with environmental impact

- based on the work of Guillen and Grossmann (2009) -

4.8.1 Introduction

In the paper by Guillen and Grossmann (2009), there is discussion of an emerging process engineering technique, known as Enterprise Wide Optimisation (EWO), whose main goal is the optimal integration of the supply, manufacturing and distribution activities in a supply chain environment. This is opposed to the current Supply Chain Management (SCM) ideology, which focuses more on the manufacturing stage. This is also in accordance with a recent trend of developing more sustainable processes since the type of analyses performed with regard to SCM/EWO analyses allows for any environmental impact to be determined. More specifically, the aim of this paper is to provide for a modelling and computational framework that can quantify the effect of a certain type of environmental uncertainty has on a supply chain model.

4.8.2 Problem Description

The supply chain environment concerned consists of plants, distribution centres with associated customer zones. There are also fixed and variable costs associated with capacity extensions of plants and distribution centres, each with upper and lower bound capacity limits. Product prices and market demand information is also known. The main objective is to determine the optimal network configuration as well as all the associated planning decisions with the intention of maximising NPV and minimising environmental impact.

Table 4-31: Nomenclature for the Pareto optimisation of a chemical supply chain -Guillen and Grossmann (2009)

Indices	Physical Meaning
b	environmental burdens
c	impact categories
d	damage categories
i	manufacturing technologies
j	plants
k	warehouses
l	markets
p	products
t	time periods
Sets	Physical Meaning
ID(d)	set of impacts c contributing to damage category d
IN(p)	set of manufacturing technologies that consume p
MP(i)	set of main products p of technology i
OUT(p)	set of manufacturing technologies that produce p
Parameters	Physical Meaning
* $CE^{PL}{}_{jt}$	capacity expansion of manufacturing technology i at plant j in time period t
* $CE^{WH}{}_{ijt}$	capacity expansion of a warehouse k at plant j in time period t
* $D_{ijt}{}^{MK}$	demand of product p sold at market i in period t
ir	interest rate
FCl	total capital investment
$NEXP_{ij}{}^{PL}$	Maximum number of capacity expansions for technology i available at plant j
$NEXP_{ik}{}^{WH}$	Maximum number of capacity expansions for technology i available for warehouse k
NT	No. of time periods
* PU_{jpt}	purchase of product p at plant j in period t
* $Q_{jkt}{}^{PL}$	flow of materials between plant j and warehouse k in period t
* $Q_{klt}{}^{WH}$	flow of materials between warehouse k and market l in period t
SV	salvage value
TQR_k	turnover ratio of warehouse k

μ_{ip}	mass balance coefficient associated with product *p* and manufacturing technology *i*
ϕ	tax rate
γ_{lpt}^{FP}	price of final product *p* sold at market *l* in time period *t*
γ_{jpt}^{RM}	price of raw material *p* sold at plant *j* in time period *t*
υ_{jt}	operating cost of manufacturing technology *i* available at plant *p* in time period *t*
π_{kt}	inventory cost at warehouse *k* in period t
Ψ_{pjkt}^{PL}	unit transportation of product *p* sent from plant *j* to warehouse k in time period t
Ψ_{klpt}^{WH}	unit transportation of product *p* sent from warehouse *k* to market *l* in time period *t*
α_{ijt}^{PL}	variable investment term associated with technology *i* at plant *j* in time period *t*
α_{kt}^{WH}	variable investment term associated with warehouse *k* in time period *t*
β_{ijt}^{PL}	fixed investment term associated with technology *i* at plant *j* in time period *t*
β_{kt}^{WH}	fixed investment term associated with warehouse *k* in time period *t*
β_{jkt}^{TPL}	fixed investment term associated with the establishment of a transport link between plant *j* and warehouse *k* in time period *t*
β_{klt}^{TWH}	fixed investment term associated with the establishment of a transport link between warehouse *k* and market *l* in time period *t*
w_{bp}^{PU}	emissions/feedstock requirement of chemical b per unit of raw material p generated
w_{bp}^{PR}	emissions/feedstock requirement of chemical b per unit of final product p generated
w_b^{EN}	emissions/feedstock requirement of chemical b per unit of energy consumed
w_b^{TR}	emissions/feedstock requirement of chemical b per unit of mass transported one unit distance
η_{ijp}^{EN}	energy consumed per unit of chemical *p* produced with manufacturing technology *i* at plant *j*
λ_{jk}^{PL}	distance between plant *j* and warehouse *k*
λ_{kl}^{WH}	distance between warehouse *k* and market *l*
θ_{bc}	damage factor of chemical *b* contributing to impact category *c*
σ_{bc}^{DF}	standard deviation of damage factor of chemical *b* contributing to impact category *c*
Ω_d	target level for damage category *d*
K	joint probability of not exceeding the individual target levels *d*
ζ_d	normalization factor for damage category *d*

ε_d	weighting factor for damage category *d*
τ	minimum percentage of available capacity that must be utilised
Variables	Physical Meaning
C_{ijt}^{PL}	capacity of manufacturing technology at plant *i* in period *t*
CE_{ijt}^{PL}	capacity expansion of manufacturing technology at plant *i* in period *t*
C_{kt}^{WH}	capacity of warehouse k in time *t*
CE_{kt}^{WH}	capacity expansion of warehouse *k* in time *t*
CF_t	cash flow in period *t*
DAM_d	impact in damage category *d*
$D\hat{A}M_d$	mean value in damage category *d*
DAM_d^{SD}	standard deviation of impact in danger category *d*
DEP_t	depreciation term in period *t*
ECO_{99}	Eco-indicator
FCI	Fixed capital investment
$FTDC_t$	Fraction of total depreciable capital paid in period *t*
IL_{kt}	average inventory level at warehouse *k* in period *t*
IM_c	damage in impact category *c*
INV_{kpt}	inventory of product *p* kept at warehouse *k* in period *t*
LCI_b	life cycle inventory entry associated with chemical *b*
NE_t	net earnings in in period *t*
NPV	net present value
PU_{jpt}	purchases of product *p* made by plant *j* in period *t*
Q_{jkpt}^{PL}	flow of product *p* from plant to warehouse *k* in period *t*
Q_{klpt}^{WH}	flow of product *p* from warehouse *k* to market *l* in period *t*
SA_{lpt}	sales of product *p* at market *l* in period *t*
W_{ijpt}	input/output flow of product *p* of tech. *i* at plant *l*
X_{ijt}^{WH}	binary variable (1, if capacity of manuf. tech. *i* at plant *j* is expanded to in period *t*, 0 otherwise)
X_{kt}^{WH}	binary variable (1, if capacity of warehouse *k* is expanded in period *t*, 0 otherwise)
Y_{jkt}^{PL}	binary variable (1, if a transportation link between plant *j* and warehouse k is established in period *t*, 0 otherwise)
Y_{klt}^{WH}	binary variable (1, if a transportation link between warehouse *k* and market *l* is established in period *t*, 0 otherwise)

Note: *When a variable is underlined, i.e.,* $\underline{Q_{klt}^{WH}}$, *it indicates a lower bound limit and when it is struck through, it indicates an upper bound limit, i.e.* ~~PU_{jpt}~~

4.8.3 Program Formulation

Further to the work of Guillen and Grossmann (2009), the mathematical formulation of the MINLP is as follows:

4.8.3.1 Mass Balances

The mass balances for the manufacturing plants and warehouses are expressed as constraints as follows:

$$PU_{jpt} + \Sigma_{i=OUT(p)} W_{ijt} = \Sigma_k Q_{jkpt}{}^{PL} + \Sigma_{i \in IN(p)} W_{ijt} \qquad \forall j, p, t \quad (1)$$

$$W_{ijt} = \mu_{ip} W_{ijt} \qquad \forall j, p, t \quad (2)$$

$$INV_{kpt-1} + \Sigma_j Q_{jkpt}{}^{PL} = \Sigma_t Q_{klt}{}^{WHU} + INV_{kpt} \qquad \forall j, p, t \quad (3)$$

Constraints (4) and (5) reflect the imposition of upper, $\overline{PU}_{jpt}$, and lower $\underline{PU}_{jpt}$, limits on the purchases of raw materials, and, and likewise on the sales of products, $\overline{SA}_{lpt}$ and SA_{lpt}. The sales of products are determined from the flows of materials between the warehouses and the final markets (6).

$$\underline{PU}_{jpt} \leq PU_{jpt} \leq \overline{PU}_{jpt} \qquad \forall j, p, t \quad (4)$$

$$\underline{D_{ijt}{}^{MK}} \leq SA_{lpt} \leq \overline{D}_{ijt}{}^{MK} \qquad \forall j, p, t \quad (5)$$

$$\Sigma_t Q_{klt}{}^{WH} = SA_{lpt}, \qquad \forall j, p, t \quad (6)$$

4.8.3.2 Capacity Constraints

a) Plant capacity constraints

Eqns. (7), (8), (9) and (10) describe the capacity constraints associated with the various production plant

$$\underline{CE_{ijt}{}^{PL}} < CE_{ijt}{}^{PL} < \overline{CE}_{ijt}{}^{PL} \quad (7)$$

$$C_{ijt}{}^{PL} = C_{ijt-1}{}^{PL} + CE_{ijt}{}^{PL} \quad (8)$$

$$\Sigma\, X_{kt}{}^{WH} \leq NEXP_{ij}{}^{PL} \quad (9)$$

$$\tau C_{ijt}{}^{PL} \leq W_{ijt} \leq C_{ijt}{}^{PL} \quad (10)$$

b) Warehouse constraints

The warehouse constraints, (11) – (13), represent the total capacity, C_{ijt}^{PL}, and capacity expansion, CE_{ijt}^{PL}, associated with warehouses in period *t.*

$$\underline{CE^{WH}}_{ijt} \leq CE^{WH}_{ijt} \leq \overline{CE^{WH}}_{ijt} \qquad \forall j, t \qquad (11)$$

$$C_{kt}^{WH} = C_{kt-1}^{WH} + CE_{kt}^{WH} \qquad (12)$$

$$\sum X_{kt}^{WH} \leq NEXP_{ik}^{WH}, \qquad \forall k, t \qquad (13)$$

c) Inventory constraints

Equations (14) – (16) represent limits on warehouse inventory levels whilst eqn. (16) equates to average inventory level.

$$\Sigma_p \mathrm{INV}_{kpt} \leq C_{kt}^{WH}, \qquad (14)$$

$$2IL_{kt} \leq C_{kt}^{WH}, \qquad \forall k, t \qquad (15)$$

$$IL_{kt} = \Sigma_t \Sigma_p Q_{klpt}^{WH} / \; TQR_k \qquad \forall k, t \qquad (16)$$

d) Transportation constraints

Product quantities transported from plants to warehouses must also lie between upper and lower limits provided, of course, that corresponding network nodes, (17) and (18), exist.

$$\underline{Q_{jkt}}^{PL} Y_{jkt}^{PL} \leq Q_{jkt}^{PL} \leq \overline{Q_{jkt}}^{PL} \qquad \forall j, k, t \qquad (17)$$

$$Q_{klt}^{WH} Y_{klt}^{WH} \leq Q_{klt}^{WH} \leq \overline{Q_{klt}}^{WH} Y_{klt}^{WH} \qquad \forall j, k, IL, t \qquad (18)$$

4.8.3.3 *Objective Function*

a) Net Present Value (NPV)

NPV is determined from the cashflow, i.e. net earnings plus investment income per period as follows:

$$NPV = \sum CF_t / (1 + ir)^{t-1} \text{ where:} \quad (19)$$

$$CF_t = NE_t - FTDC_t, \quad t = NT - 1 \quad (20)$$

$$CF_t = NE_t - FTDC_t + SV.FCI, \quad t = NT \quad (21)$$

, where:

$$NE_t = (1 - \phi)[\sum_t \sum_p \gamma_{lpt}^{FP} SA_{lpt} - \sum_t \sum_p \gamma_{jpt}^{RM} PU_{jpt} - \sum_i \sum_j \sum_{MP(i)} \upsilon_{jt} W_{ijpt} - \sum_k \pi_{kt} IL_{kt} - \sum_j \sum_k \sum_p \Psi_{jkpt}^{PL} Q_{jkpt}^{PL} - \sum_k \sum_l \sum_p \Psi_{klpt}^{WH} Q_{klpt}^{PL}] + \phi DEP_t, \quad \forall k, t \quad (22)$$

$$DEP_t = (1 - SV) FCI/NT \quad \forall t \quad (23)$$

$$FCI = \sum_i \sum_j \sum_t (\alpha_{ijt}^{PL} CE_{ijt}^{PL} + \beta_{kt}^{WH} X_{kt}^{WH}) + \sum_i \sum_j \sum_t (\beta_{jkt}^{TPL} Y_{jkt}^{PL}) + \sum_k \sum_l \sum_t (\beta_{klt}^{TWH} Y_{klt}^{WH}) \quad (24)$$

$$FCI \leq \overline{FCI} \quad (25)$$

$$FTDC_t = FCI/NT \quad \forall t \quad (26)$$

b) Environmental Impact

The environmental performance of the network is measured by the ECO-Indicator 99. For this purpose, the calculation of life-cycle inventory associated with the supply chain operation, which includes the emissions released as well as the feedstock requirements associated with the manufacturing and distribution activities, is required. These life cycle inventories are associated with raw material consumption, PU_{jpt}, the manufacture of final products, W_{ijpt}, the transportation of materials between plants and warehouses, Q_{jkpt}^{PL}, and then the final transportation between warehouses and markets, Q_{klpt}^{WH}. These accumulated inventories (27) are then translated into environmental damages, using equation (28).

$$LCI_b = \sum_j \sum_p \sum_t w_{bp}^{PU} PU_{jpt} + \sum_i \sum_j \sum_{MP(i)} \sum_t w_{bp}^{PR} W_{ijpt} - \sum_i \sum_j \sum_{MP(i)} \sum_t w_b^{EN} \eta_{ijp}^{EN} W_{ijpt} - \sum_j \sum_k \sum_p \sum_t w_b^{TR} \lambda_{jk}^{PL} Q_{jkpt}^{PL} - \sum_k \sum_l \sum_p \sum_t w_b^{TR} \lambda_{kl}^{WH} Q_{jkpt}^{WH} \quad \forall b \qquad (27)$$

$$DAM_d = \sum_{c=ID(d)} \sum_b \theta_{bc} LCI_b \quad \forall d \qquad (28)$$

The impacts in each damage category can be accumulated into a single relationship:

$$ECO_{99} = \sum_d \varepsilon_d \zeta_d \cdot DAM_d \qquad (29)$$

4.8.4 Environmental performance under uncertainty

A joint chance-constraint is applied in this model to explicitly describe the environmental performance under uncertainty. This probabilistic constraint imposes a probability target for simultaneously satisfying the desired environmental requirement in each damage category. Such an approach has the advantage of providing simultaneous control over a set of stochastic environmental variables, without which would be of concern to decision makers.

Life cycle inventories are assumed to be known in this model, whereas the damage factors, θ_b, will be described through Gaussian probability functions and for this purpose, the widely used normal probability distribution function will be applied.

The aforementioned joint chance-constraint can be expressed as follows:

$$\Pr[\cap DAM_i \leq \Omega_d] \geq K \qquad (30)$$

, where K represents the lower limit for joint probability of not exceeding the individual target levels, Ω_d, that are defined for each damage category, DAM_i. The value of K, which gives an indication of environmental performance under uncertainty, can then be applied to the objective function as an additional criterion to be maximised along with the NPV. If such damage factors are assumed to be independent random variables then the scope of joint chance-constraint can be extended as follows:

$$\prod_{d} \Pr\,[\mathrm{DAM}_{i} \leq \Omega_{d}] \geq K \tag{31}$$

If it is further assumed that the damage factors follow normal distributions, then the following relationships will apply

$$\prod_{d} \Pr \left\{ \frac{DAM_{d} - D\hat{A}M_{d}}{DAM_{d}^{SD}} < \frac{\Omega_{d} - D\hat{A}M_{d}}{DAM_{d}^{SD}} \right\} \tag{32}$$

$$\prod_{d} \Pr\, \phi \left\{ \frac{\Omega_{d} - D\hat{A}M_{d}}{DAM_{d}^{SD}} \right\} \tag{33}$$

The mean and standard deviation of the impact caused in each category are calculated from the inventory life cycle and damage factors as follows

$$DAM_{d}^{SD} = \left\{ \Sigma_{c \in ID(d)}\, \Sigma_{b} (\sigma_{bc}^{DF}\, LCI_{b})^{2} \right\} \qquad \forall d \tag{34}$$

$$D\hat{A}M_{d} = \left\{ \Sigma_{c \in ID(d)}\, \Sigma_{b} \theta_{bc}\, LCI_{b} \right\} \qquad \forall d \tag{35}$$

A logarithmic transformation is applied to constraint (33) in order to express the product as a summation of logarithmic functions

$$\Sigma_{d} \ln \phi \left\{ \frac{\Omega_{d} - D\hat{A}M_{d}}{DAM_{d}^{SD}} \right\} > \ln K \tag{36}$$

Consequently, the bi-criterion MINLP may be generated:

$$\sum_{d} \ln\phi \left\{ \frac{\Omega_d - D\hat{A}M_d}{DAM_d^{SD}} \right\} \text{ (P1): } \max_{x,y} \ \text{NPV(x, y)},$$

, subject to eqns. (1) – (27) and (34) & (35)

Model (P!) can then be solved by any standard algorithm for a multi-objective optimisation program, and in order to approximate a Pareto set, the ε-constraint method of Chankong and Haimes (1979), is proposed that must be solved for different values of the auxiliary parameter, $\in$.

$$\sum_{d} \ln\phi \left\{ \frac{\Omega_d - D\hat{A}M_d}{DAM_d^{SD}} \right\} \text{ (P2): ma}$$

subject to NPV(x, y)

eqns. (1) – (27) and (34) & (35)

The objective function (P2) can be reformulated as follows:

$$\max \Sigma_d \ln\phi \ (A_d/B_d) \tag{37}$$

, subject to: NPV(x, y)

eqns. (1) – (27) and (37) & (35)

, where $A_d = \ \Omega_d - D\hat{A}M_d$

, and $B_d = \ DAM_d^{SD}$

The lower and upper limits that define the interval within which the epsilon, ε, parameter must fall can be obtained by solving each objective global optimally separately, i.e.:

$$\left.\begin{array}{l}(x, y) = \text{argmax}_{x,y} \Sigma_d \ln\phi\,(A_d/B_d) \\ \text{s.t. eqns. (1) – (27), (35) \& (37)}\end{array}\right\} \qquad \text{(P3a)}$$

and:

$$\left.\begin{array}{l}(x, y) = \text{argmax}_{x,y} \text{NPV}(x, \text{y}) \\ \text{s.t. eqns. (1) – (27), (35) \& (37)}\end{array}\right\} \qquad \text{(P3b)}$$

Such transformations have enabled the reformulation of (P1), the original Pareto set, into a single-objective problem, (P2) and from there into an equivalent form (P3) that has a set of constraints that define a non-convex search space.

4.8.5 Extraction of the Global Pareto-Optimisation under Multi-objective Stochastic Fuzzy Uncertainty Technique

Table 4-32: Global Pareto-Optimisation of an environmentally conscious supply chain under a Multi-objective Stochastic Fuzzy Uncertainty Technique - based on the work of Guillen and Grossmann (2009)

1 Problem description of the supply chain network including supply, production, warehousing, distribution and sales

2 Requirements Definition: Determine the optimal configuration of the supply chain network that will maximise NPV and also minimise the environmental impact

4 Define type of Uncertainty involved and Solution approach:
Fuzzy uncertainty is associated with the supply chain (production and distribution) aspects whilst stochastic uncertainty is associated with the environmental impact. These two aspects will be combined in a Pareto set to give a bi-criterion non-convex MINLP.

5 Formulate a standard MINLP, i.e. objective function plus constraints, for the determination of NPV in terms of the production and distribution aspects

6 Formulate an environmental impact MINLP, which is measured by the Eco-indicator 99 in terms of life cycle inventories associated with the emissions released and feedstock, manufacturing and distribution requirements

5 Integrate the environmental impact model with the production and distribution NPV MINLP to create a bi-criterion non-convex MINLP

6 Solution of the bi-criterion non-convex MINLP
The global maximum of the bi-criterion MINLP will provide a Pareto solution that can be generated by application of the $\in$-constraint method by providing different instances of the MINLP and also by providing bound limits. This involves a 4-step algorithm that repeatedly calculates lower and upper bound limits until convergence occurs

7 Record the global optimum solution of the bi-criterion non-convex MINLP

4.8.6 Conclusion

This paper by Guillen and Grossmann (2010) involved the optimal design and operation of a chemical supply chain, but also taking into account certain environmental impact concerns, whose effect was assessed through the application of environment impact measuring technology, Eco-indicator 99. The overall problem was formulated as a bi-criterion chance constrained MINLP with two objective functions, one of which was to maximise NPV and the other of which was to simultaneously satisfy all the environmental targets. The resulting bi-criterion non-convex MINLP was solved by application of the ε-constraint method to guarantee a global optimum Pareto solution. The overall 'Global Pareto-Optimisation under Multi-objective Stochastic Fuzzy Uncertainty Technique' was extracted and is recorded in Table 4-32.

4.9 The Solution of Multi-Objective Linear Programs through Fuzzy Goal Programming

- based on the work of Baky (2010) -

4.9.1 Introduction

The traditional approach towards the solution of mathematical linear programs only involved the inclusion of one performance objective, and also at only one decision-making level, in the overall programming mix. However, since many operational situations frequently involve the consideration of many, and often conflicting, performance objectives and this too in a hierarchical decision-making environment, the concept of Multi-Level Multi-Objective Linear Programming was introduced to account for such situations. Initial focus in this regard involved the introduction of bi-level multi-objective linear programming, to account for two decision-making levels, and this was soon supplemented by the introduction of three-level programming (TLP). However, it was soon realised that, in a hierarchical, multi-functional environment, typical of large production organisations, multiple performance objectives at multiple decision-making levels had also to be provided for. In this paper, it is therefore the intention to formulate such multi-level multi-objective linear programming (ML-MOLP) capability.

However, since the focus of this research effort is only on singular level, multi-objective programming, the concluding problem example will only deal with a singular level, multi-objective programming case study

4.9.2 Problem Description

A p-level multi-objective minimisation-type problem is considered wherein $F_i\,(x_i)$ is a suite of objective functions, of which $f_{i,j}(x_{i,j})$ are the vectors of each objective function. Therefore the minimisation problem may be expressed as:

$$\text{Min}_{x1}\, F_1(x_{1,j}) = \text{Min}_{x1}\, [f_{11}(x_{11}), f_{12}(x_{12})\ldots, f_{1m}(x_{1m})]$$

$$\text{Min}_{x2}\, F_2(x_{2,j}) = \text{Min}_{x2}\, [f_{21}(x_{21}), f_{21}(x_{21})\ldots, f_{2m}(x_{2m})]$$

$$: \quad : \quad : \quad : \quad : \quad :$$

$$\text{Min}_{xp}\, F_p(x_{p,j}) = \text{Min}_{xp}\, [f_{p1}(x_{p1}), f_{12}(x_{p2})\ldots, f_{1m}(x_{pm})]$$

subject to:

$$x \,\varepsilon\, G = \{x \,\varepsilon\, R^n | A_1x_1 + A_2x_2 + \ldots + A_px_p\ (\leq, =, \geq)\ b_i,\ x_{i,j} \geq 0,\ b \,\varepsilon\, R^n\}$$

, where G is the multi-level constraints set and where;

$$f_{ij}(x_i) = c^j_1x_1 + c^j_2x_2 + .. + c^j_px_p,\ i = 1, 2, \ldots, p,\ j = 1, 2, \ldots, m$$

$$= c^j_{11}x_{11} + \ldots + c^j_{1n}x_{1n} + c^j_{21}x_{21} + \ldots + c^j_{2n}x_{1n} + c^j_{p1}x_{p1} + \ldots + c^j_{pm}x_{pm}$$

4.9.3 Fuzzy Goal Programming

With multi-level multi-objective linear programming (ML-MOLP), imprecise Aspiration Levels (λ) are assigned to each objective function and are characterised by their associated membership functions for the achievement of those aspiration levels. Now, since there is a requirement to minimise all objective functions, the optimum minimum value of each independently can be taken as the aspiration levels of their associated fuzzy goals. Further to the work of Baky (2010), the procedure is as follows:

Let x^j_i ($i = 1\ldots p$ (no. of decision-makers/objective functions), $j = 1\ldots m$ (no. of decision-making levels)) be the optimal solutions of the corresponding objective functions when calculated in isolation and let $g_i \geq f_{ij}^{min}$ be the aspiration level assigned to the ij^{th} objective. Also let $x^p{}_i = x^p{}_1, x^p{}_2\ldots x^p{}_p$ be the optimal solution of the ith level MILP problem. Consequently, the fuzzy goals of the objective functions at each level, and the vectors of such fuzzy goals, in terms of the decision variables, re expressed as:

$f_{i,j}(x_{i,j}) <=> g_j,\ i = 1, 2, \ldots, p;\ j = 1, 2, \ldots m$

, where '<=>' indicates the fuzziness of the aspiration levels

Membership functions, $\mu_f f_{i,j}(x_{i,j})$ can consequently be formulated, as graphically represented in Fig 4-15

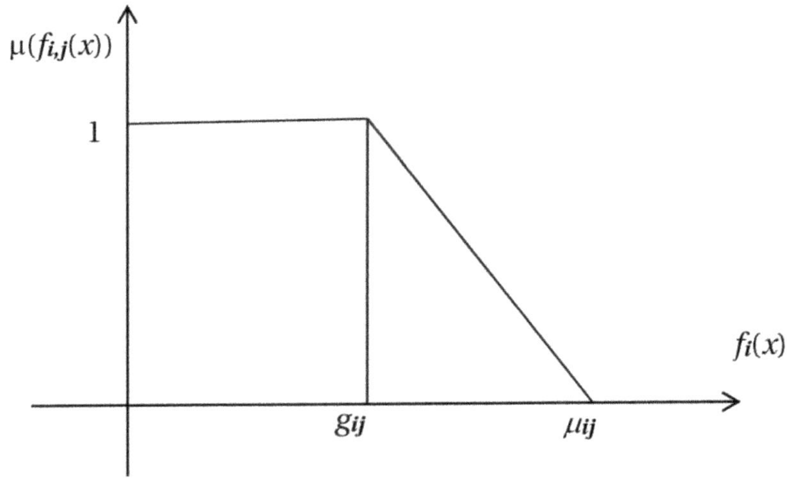

Figure 4-15: Membership function of minimisation-type objective function

$$\mu_f f_{i,j}(x_{i,j}) = \begin{cases} 1, & \text{if } f_{i,j}(x_{i,j}) \le g_{ij} \\ \dfrac{\mu_{ij} - f_{i,j}(x_{i,j})}{\mu_{ij} - g_{ij}}, & \text{if } g_{ij} \le f_{i,j}(x_{i,j}) \le \mu_{ij} \\ 0, & \text{if } f_{i,j}(x_{i,j}) \ge \mu_{ij} \end{cases}$$

Now, to build the membership functions, corresponding to the fuzzy goals of the decision variables, the optimal values of the corresponding objective functions must first be determined, i.e. $x^p = (x_1{}^p, x_2{}^p \ldots x_p{}^p)$. Let $t_k{}^{iL}$ and $t_k{}^{iR}$, k = 1, 2...n be the maximum negative and positive tolerance values on the decision vectors. Therefore each of the various linear membership functions, corresponding to the n components of the decision vector, as graphically represented in Fig 4-16, may be formulated as follows:

$$\mu_{ff_{i,j}}(x_{i,j}) = \begin{cases} \dfrac{x_{ik} - (x^*_{ik} - t_k{}^{iL})}{t_k{}^{iL}} & \text{if } x^*_{ik} - t_k{}^{iL} \leq x_{ik} \leq x^*_{ik} \\ \dfrac{(x^*_{ik} + t_k{}^{iR}) - x^*_{ik}}{t_k{}^{iR}} & \text{if } x^*_{ik} \leq x_{ik} \leq x^*_{ik} + t_k{}^{iR} \\ \mu(x_{i,k}) & \\ 0, & \text{otherwise} \end{cases}$$

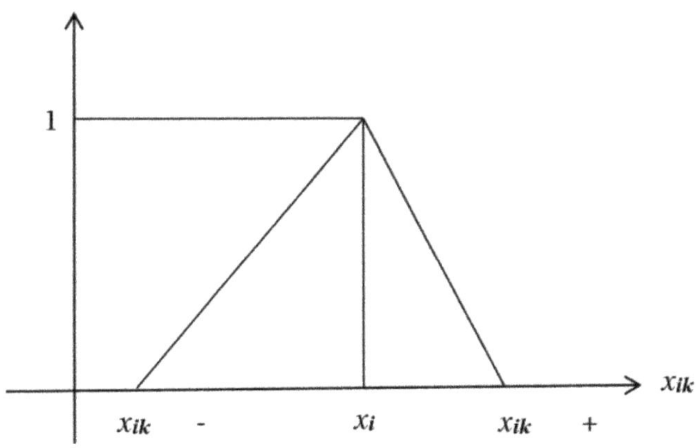

Figure 4-16: Membership functions of decision vectors

4.9.4 Fuzzy Goal Programming Algorithm for ML-MOLP's

The following algorithm, put forward by Baky (2010), solves a Multi-level Multi-Objective Linear Programming (ML-MOLP) model that describes an ith level decision-making environment

1) Calculate the individual minimum and maximum values of all the objective functions for all levels under the given constraints

2) Set the goals and the upper tolerance limits - u_{ij}, g_{jj}, $i = 1, 2 \ldots p; j = 1, 2 \ldots m$

3) Evaluate the weights, $w_{ij}^{+} = 1/(u_{iij} - g_{ijj})$, $i = 1, 2 \ldots p; j = 1, 2 \ldots m$

4) Set $l = 1$

5) Elicit the membership functions, $\mu_{f}f_{i,j}(x_{i,j})$, $j = 1, 2 \ldots m$

6) Formulate the model for the ith level MOLP problem

7) Solve the model to get: $x^{l*} = x_1^{l*} + x_2^{*} \ldots x_p^{*}$.

8) Set the maximum negative and positive tolerance levels on the decision vector, $x_l^{l*} = x_1^{l*} + x_{l2}^{*} \ldots x_{lp}^{*}$.

9) Evaluate the weights, $w_{lk}^{L} = 1/t_k^{lL}$ and $w_{lk}^{R} = 1/t_k^{lR}$, $k = 1, 2 \ldots n$

10) Elicit the membership functions: $\mu_{f(pj)}(f_{f(pj)}(x))$, $j = 1, 2 \ldots m$

11) $l = l + 1$

12) If $l > p - 1$, then go to step 13), otherwise go step 5.

13) Elicit the membership functions, $\mu_{f}f_{i,j}(x_{i,j})$, $j = 1, 2 \ldots m$ for the objective functions in the pth level

14) Evaluate the weights $w_{ij}^{+} = 1/(u_{iij} - g_{ijj})$, $j = 1, 2 \ldots m$

15) Formulate the model for the ML-MOLP problem

16) Solve the model to get the optimum solution for the ML-MOLP problem

4.9.5 Numerical Example

Problem:

Consider the following i^{th} level ML-MOLP problem;

Min_{x1} ($f_{11} = x_1 - x_2 - 4x_3$, $f_{11} = -x_1 + 3x_2 - 4x_3$)

, subject to;

$x_1 + x_2 + x_3 \leq 3$, $x_1 + x_2 - x_3 \leq 1$

$x_1 + x_2 + x_3 \geq 1$, $-x_1 + x_2 + x_3 \leq 1$

$x_1 \leq 0.5$ and $x_1, x_2, x_3 \geq 0$

Solution Procedure:

1) Calculate the individual minimum and maximum values of all the objective functions for all levels under the given constraints

2) Set the goals and the upper tolerance limits - u_{ij}, g_{jj}, $i = 1, 2 \ldots p; j = 1, 2 \ldots m$

3) Evaluate the weights, $w_{ij}^{+} = 1/(u_{ij} - g_{ij})$, $i = 1, 2 \ldots p;$

4) Elicit the membership functions, $_{f}f_{i,}(x)$, $i = 1, 2 \ldots p$

$$5)\ \ {}_{i}f(x_{i}) = \begin{cases} 1 \\ ({}_{I} - f(x_{i}))/(u_{I} - g_{i}) \\ 0 \end{cases}$$

5) Formulate the model for the ith level MOLP problem

6) Solve the model, min ($f(x_i)$) to obtain the optimum variables, $x^{l*} = x_1^{l*} + x_2^{*} \ldots x_p^{*}$.

7) Calculate the values of the various corresponding objective functions

Objective Functions:

$\text{Min}_x(f_{11} = x_1 - x_2 - 4x_3)$

$\text{Min}_x(f_{12} = -x_1 + 3x_2 - 4x_3)$

Variables	u_i		g_i	x_1	x_2	x_3
Values	$f_{11}(x)$		$f_{12}(x)$	0	1	0
1st Level	Min		Max			
Min_{x1} ($f_{11} = x_1 - x_2 - 4x_3$)	-2.5	-1	1	1	-1	-4
Min_{x1} ($f_{12} = -x_1 + 3x_2 - 4x_3$)	-3.5	3	3	-1	3	-4
(*) $Max_{1(x)} = 0.714 - 0.286x_1 + 0.286x_2 + 1.143x_3$	1		1.429	-0.286	0.286	1.143
(*) $Max_{2(x)} = 0.538 + 0.154x_1 - 0.462x_2 + 0.615x_3$	0.076		1.0765	0.154	-0.462	0.615
Constraints						
$x_1 + x_2 + x_3 < 3$	1		3	1	1	1
$x_1 + x_2 - x_3 < 1$	1		1	1	1	-1
$-x_1 + x_2 + x_3 < 1$	1		1	-1	1	1
$x_3 < 0.5$	0		0.5			1
$x_1 + x_2 + x_3 > 1$	1		1	1	1	1
$x_1 > 0$	0		0	1		
$x_2 > 0$	1		0		1	
$x_3 > 0$	0		0			1

(*): $f(x) = (u_1 - f(x_1))/(u_1 - g_1)$

Now $(u_1, g_1) = (-2.5, 1)$

$_1f(x_1) = (-2.5 - (x_1 - x_2 - 4x_3))/(-2.5 - 1)$

$= (-2.5 - (x_1 - x_2 - 4x_3))/(-2.5 - 1)$

$= 0.714 - 0.286x_1 + 0.286x_2 + 1.143x_3$

Max ($_1f(x_1)$) = $_1$ = 1..4285

Similarly, $f(x) = (u_2 - f(x_2))/(u_2 - g_2)$

Now $(u_2, g_2) = (-3.5, 3)$

$_2f(x_2) = (-3.5 - (-x_1 + 3x_2 - 4x_3))/(-3.5 - 3)$

$= (-3.5 + x_1 - 3x_2 + 4x_3))/(-6.5)$

$= 0.538 - 0.154x_1 + 0.462x_2 - 0.615x_3$

, and Max ($_2f(x_2)$) = $_2$ = 1.0765

Now Opt ($f(x)$) = min ($_1$, $_2$) = min (1.4285, 1.0765) = 1.0765

$f_{11}^{min}(opt) = -0.5$ and $f_{12}^{min}(opt) = -3.5$ $\Longrightarrow$

4.9.6 Extraction of the optimisation procedure for a multi-objective problem under stochastic fuzzy uncertainty

Table 4-33: Optimisation procedure for a bi-objective supply chain network under stochastic fuzzy uncertainty based on the work of Baky (2010)

1	**Description of the problem**
2	Requirements Definition: Minimise multiple objective functions
3	Nomenclature Requirements
4	Define type of Uncertainty involved and Solution approach: Fuzzy and multi-objective uncertainty is associated with the problem. The solution procedure is to transform the objective functions into membership functions and then solve the corresponding MILP model
5	Elicit the membership functions per objective function, $f_j(x)$ - Set the goals and limits, u_j and g_j per obj. fn. - Elicit the membership fns., $\mu_i(x) = [u_j - f_i(x)] / [u_j - g_j]0^1$
6	Transform the membership function into a MILP program with corresponding objective functions and constraints
5	Solve the MILP per objective
6	Record the optimum results

4.9.7 Conclusion

This paper presented procedures for solving multi-level multi-objective programming problems although, since the focus of this current research effort is on singular level multi-objective programming, only those aspects of his paper were focussed on in the problem example. This was achieved by minimising the group regret of degree of satisfaction of the decision makers in order to achieve the highest degree (unity) of each of the defined membership functions to the extent possible by minimising their deviational variables, and thereby obtaining the most satisfactory solution for each of the decision makers.

The supply chain, under uncertainty, optimisation technique utilised with this procedure was extracted and is documented in Table 4-33.

4.10 The Optimisation of a Bi-objective Supply Chain under Fuzzy uncertainty

- based on the work of Cadona-Valdés and Ozdemir (2011) -

4.10.1 Introduction

In the paper by Cadona-Valdés and Ozdemir (2011) wherein the design of a bi-objective supply chain network is discussed, the focus of the research is on distribution facility population and location, which is considered important in defining the distribution network of the supply chain. This is because, although new distribution facilities incur both establishment and additional inventory costs, they also reduce the time to market and thus provide competitive advantage.

In this work, all the cost implications of establishing new distribution centres are considered, which include the fixed costs of establishment as well as the effect on transportation nature and costs. Therefore, the intention is to minimise total cost as well as the total shipping time across the network. Consequently, the contribution of this paper is twofold: Firstly, greater operational realism is considered in the study by applying a bi-objective stochastic mixed integer linear programming approach and secondly, the solution approach is based on a fusion of the ε-constraint and L-shaped techniques.

4.10.2 Problem Description

Several, different transport alternatives, between the various different supply chain locations, are considered in this paper on account of the existence of many different options available in terms of cost and time parameters. Therefore, the objective of this work is to determine the optimum distribution configuration by minimising the total logistics costs, involving both infrastructure and transportation costs, whilst taking into account customer demand uncertainty and customer satisfaction levels, in terms of lead times. This trade-off between cost and time give rise to a bi-objective MINLP problem. Although previous research had considered this cost-time trade-off before, it did not relate this balance to the availability of different inter-facility transport options.

Such cost-time trade-off, in conjunction with uncertain market demand implies a novel multi-objective stochastic optimisation problem.

The supply chain network involved is a two-echelon system consisting of manufacturing plants, distribution centres and customers. There are also a number of possible locations for additional distribution centres, for which both the establishment costs and the limited capacities are known. Different transport options are also available between the various facilities. Market demand is uncertain and is considered a random variable, which is modelled through various probability scenarios. The intention is to achieve a single optimum network design that will be identical for every demand scenario, and the decisions that need to be taken include the number and location of distribution centres to be opened, the selection of transportation channels and the product flow between facilities to meet customer demand under the various different demand scenarios.

The nomenclature utilised for the program modelling is given in Table 4-34

Table 4-34: Nomenclature for bi-objective supply chain modelling - Cardona-Valdés and Ozdemir (2011)

Sets	Physical Meaning
I	set of plants, i
J	set of potential locations, j, for distribution centres
K	set of customers, k
LP_{ij}	set of arcs, l, between nodes i and j
LW_{ij}	set of arcs, l, between nodes j and k
S	no. of demand scenarios
Parameters	Physical Meaning
CP_{ijl}	cost of transporting one unit of product from plant i to distribution centre j using arc l
CW_{jkl}	cost of sending one unit of product from plant i to distribution centre j to customer k using arc l
TP_{ijl}	time for transporting any quantity of product from plant i to distribution centre j using arc l
TW_{jkl}	time for transporting any quantity of product from plant i to distribution centre j using arc l
MP_i	capacity of plant i
MW_j	capacity of distribution centre j
F_j	fixed cost for opening distribution centre j
D^s_k	demand at customer k in scenario s
p_s	probability of realisation of scenario s
Decision Variables	Physical Meaning
X^s_{ijl}	Quantity transported from plant i to DC j in scenario s using arc l
Y^s_{jkl}	Quantity transported from DC j to customer k in scenario s using arc l
Binary Variables	Physical Meaning
Z_j	binary variable equal to 1 if DC j is open and equal to 0, otherwise
A_{ijl}	binary variable equal to 1 if arc l connecting plant i to DC j is used, and 0 otherwise
B_{jkl}	binary variable equal to 1 if arc l connecting DC j and customer k is used, and 0 otherwise

4.10.3 Trade-off between network cost and network response time

Since the intention of this paper is to determine the optimum trade-off between network cost and network response time in order to determine an optimum distribution strategy in terms of the number and capacity of distribution centres, a corresponding bi-objective stochastic mixed integer linear program model, which has both an economic and service quality perspective, is formulated, according to Cardona-Valdés and Ozdemir (2011) as follows;

$$\min f_1 = \sum_{jkl} F_j Z_j + \sum_{s \in S} p_S \left(\sum_{i \in I} \sum_{j \in J} \sum_{k \in K} CP_{ijl} X^s_{ijl} + \sum_{j \in J} \sum_{k \in K} \sum_{l \in LWjk} CW_{jkl} Y^s_{jkl} \right) \quad (1)$$

The objective that minimises the sum of the maximum lead times from the plants to the customers through each distribution centre is expressed as follows:

$$\min f_2 = \max_j \left(\max_{il} \left(TP_{ijl} A_{ijl} \right) + \max_{kl} \left(TW_{jkl} B_{jkl} \right) \right) \quad (2)$$

The establishment of the following constraint is necessary to guarantee customer satisfaction in any possible scenario:

$$\sum_{j \in J} \sum_{l \in LWjk} Y^s_{jkl} = D^S_k, \ \forall s \in S, \ k \in K \quad (3)$$

The following constraints are imposed for not exceeding plant capacity limits:

$$\sum_{j \in J} \sum_{l \in LWjk} X^s_{jkl} = MP_i, \ \forall s \in S, \ i \in I \quad (4)$$

The following constraints guarantee the desired performance of the distribution centres, i.e. the net incoming flow does not exceed the capacity of the distribution centres, (5), the flow balance at each DC is maintained, (6), that each DC can only be supplied and guaranteed from one source, (7), (8), (9) and that DC is not open if no incident arcs are involved, (10)

$$\mathrm{MW}_j Z_j - \sum_{k\in K}\sum_{l\in LWjk} \mathrm{Y^s}_{jkl} \geq 0,\ \forall s \in S, j \in J \qquad (5)$$

$$\sum_{i\in I}\sum_{l\in LWjk} \mathrm{X^s}_{ijl} - \sum_{j\in J}\sum_{l\in LWjk} \mathrm{Y^s}_{jkl} = 0,\ \forall s \in S, j \in J \qquad (6)$$

$$\sum_{j\in J}\sum_{l\in LWjk} \mathrm{B}_{jkl} = 1,\ \forall k \in K \qquad (7)$$

$$\sum_{l\in LWjk} \mathrm{A}_{ijl} \leq 1,\ \forall i \in I, j \in J \qquad (8)$$

$$\sum_{l\in LWjk} \mathrm{B}_{jkl} \leq 1,\ \forall j \in J, k\in J \qquad (9)$$

$$\sum_{i\in I}\sum_{l\in LPij} \mathrm{A}_{ijl} - \mathrm{Z}_j \geq 0,\ \forall j \in J, \qquad (10)$$

The following constraints signify a network design that is consistent for all demand scenarios, (11) and (12) and that flows are only permitted through active arcs, (13) and (14):

$$\sum_{i\in I}\sum_{l\in LWjk} (\mathrm{X^s}_{ijl} - \mathrm{A}_{ijl}) \geq 0,\ \forall i \in I, j \in J \qquad (11)$$

$$\sum_{j\in J}\sum_{l\in LWjk} (\mathrm{Y^s}_{jkl} - \mathrm{B}_{jkl}) \geq 0,\ \forall j \in J, k \in K \qquad (12)$$

$$\mathrm{MP}_j \mathrm{A}_{ijl} - \mathrm{X^s}_{ijl} \geq 0,\ \forall s \in S, j \in J, l \in LP_{ij} \qquad (13)$$

$$\mathrm{MWB}_{jkl} - \mathrm{Y^s}_{jkl} \geq 0,\ \forall s \in S, j \in J, l \in LP_{ij} \qquad (14)$$

Constraint (15) is for continuous non-negative variables and constraint (16) covers binary variables

$$\mathrm{X^s}_{ijl}, \mathrm{Y^s}_{jkl} \geq 0\ \forall s \in S, j \in J, k \in K, l_1 \in LP_{ij}, l_2 \in LW_{ij}, \qquad (15)$$

$$\mathrm{Z}_j, \mathrm{A}_{ijl}, \mathrm{B}_{jkl} \in (0, 1)\ \forall i \in I, j \in J, k \in K, l_1 \in LP_{ij}, l_2 \in LW_{ij}, \qquad (16)$$

Auxiliary variables:

- E_j^1 - maximum time available in the first echelon of the supply chain for active DC, ,*j*, i.e. $E_j^1 = \max(TP_{ijl} A_{ijl})$
- E_j^2 - maximum time available in the second echelon of the supply chain for active DC, ,*j*, i.e. $E_j^2 = \max(TW_{jkl} B_{jkl})$
- T – maximum time allowed for sending product from any plant to any distribution centre, i.e. $T = \max(E_j^1 + E_j^2)$

Auxiliary variable related constraints

$T - E_j^1 - E_j^2 \geq 0, \forall j \in J$ (17)

$E_j^1 - TP_{ijl} A_{ijl} \geq 0, \forall i \in I, j \in J, l \in LP_{ij}$ (18)

$E_j^2 - TW_{jkl} B_{jkl} \geq 0, \forall j \in J, k \in K, l \in LW_{jk}$ (19)

$T, E_j^1, E_j^2 \geq 0$ (20)

Stochastic bi-objective mixed integer linear model

$$\min f_1 = \Sigma_{jkl} F_j Z_j + \Sigma_{s \in S}\, p_S (\Sigma_{i \in I} \Sigma_{j \in J} \Sigma_{k \in K} CP_{ijl} X^s{}_{ijl} + \Sigma_{j \in J} \Sigma_{k \in K} \Sigma_{l \in LWjk} CW_{jkl} Y^s{}_{jkl})$$

$$\min f_2 = T$$

subject to constraints (3) – (20)

4.10.3.1 *Program formulation as a two-stage stochastic model*

A two-stage stochastic liner model with recourse is proposed to account for a situation wherein product network flows are dependent on the resolution of demand uncertainties, which in turn are dependent on the choice and location of distribution centres. The general formulation of a two-stage stochastic program with recourse is as follows:

$$\min (z = c^2x + E_\xi \{\min q(\omega)W = h(\omega) - T(\omega), x, y \in Y\} \quad (21)$$

subject to: $Ax = b, x \in X$

, where the first stage results are represented by the vector, x and corresponding to that are the known vectors, c, b as well as the matrix, A. The second stage results, $q(\omega)$, $h(\omega)$ and $T(\omega)$ only become known once the value of ω is realised. W is called the recourse matrix. Under such an arrangement, distribution centre locations as well as service choice for the two levels will be realised at first before the network flows can be determined. Implementation adjustments are made to the model to cater for satisfactory recourse.

4.10.4 Solution Procedure

A Pareto solution approach to this bi-objective stochastic mixed integer linear programming problem requires that the program be solved in two phases. In the first phase, the bi-objective stochastic program is transformed into a deterministic equivalent and in the second phase, multi-objective programming techniques are applied to transform the bi-objective problem into a single objective stochastic linear programming problem. The transformation of the bi-objective stochastic mixed integer linear programming problem into a single objective equivalent is achieved by application of the $\in$-constraint method, in which one objective (i.e. the principal one, if it exists) objective function is retained as the main objective function whilst the others are transformed into constraints, i.e.:

multiple objective problem, (22)

$$\min [f_1(x), \ldots, f_p(x)]$$

is substituted by the $\in$-constraint:

$$\min f_j(x)$$

subject to: $f_k(x) \leq \in_k, k = 1, \ldots, p, k \neq j$

where $\in_k = (\in_1, \in_2 \ldots \in_p)$

4.10.5 Extraction of the optimisation technique for a bi-objective supply chain under uncertainty

Table 4-35: Optimisation technique for a bi-objective supply chain under fuzzy uncertainty based on the work of Cadona-Valdés and Ozdemir (2011)

1	Problem description of the supply chain network including supply, production, warehousing, distribution and sales and transportation
2	Requirements Definition: Determine the optimal design of the supply chain network that will maximise NPV and also minimise the delivery lead time
3	Define type of Uncertainty involved and Solution approach: Fuzzy uncertainty is associated with the supply chain (production and distribution) aspects, stochastic uncertainty is associated with uncertain demand whilst multi-objective (bi-objective) uncertainty is associated with the need to minimise both the costs and the delivery lead time in a Pareto optimal fashion
4	Formulate the bi-objective stochastic mixed integer linear program
5	Transform the bi-objective stochastic mixed integer linear program into a deterministic equivalent
6	Transform the deterministic equivalent into a single objective mixed integer linear program using the Pareto optimal $\in$-constraint method
7	Solve the resulting single objective mixed integer linear program by utilising the L-shaped technique, which handles any feasibility issues arising from stochastic programming
8	Record the Pareto optimal bi-objective result

4.10.6 Conclusion

In this paper by Cardona-Valdés and Ozdemir (2011), the optimal design of a supply chain network was discussed wherein both the network cost and the network response time was minimised in a Pareto optimal fashion using the $\in$-constraint technique. A bi-objective stochastic mixed integer linear programming problem was initially developed and then, to guarantee a relatively complete recourse property in solving it as a two stage stochastic model, a fictitious distribution centre, with sufficient capacity to handle customer demand, was introduced. The L-shaped method was utilised in the solution process to handle any feasibility issue arising from stochastic programming.

The optimisation procedure utilised herein was extracted and is tabulated in Table 4-35.

4.11 Multi-Objective (Pareto-) Optimisation of Supply Chains in the process industry

- based on the work of Liu and Papageorgiou (2013) -

4.11.1 Introduction

A supply chain consists of a number of sequential operations, i.e. production, distribution and sales, that work together to transform raw materials into products and then supply them to customers. Cost of operations and customer responsiveness, as well as timeous delivery, are widely regarded as the three most important criteria in measuring the performance of supply chains, and therefore these three performance metrics are crucial for supply chain design and planning.

Most previous supply chain research models only considered a single criterion for supply chain planning and optimisation but in the last ten years, an increasing amount of research has been conducted on multi-objective problems and solution methods. Of these the ε-constraint method has been widely used to generate Pareto optimal solutions. However, since very little previous work has been done on the trade-off between supply chain cost, responsiveness and service levels, the purpose of the Liu and Papageorgiou (2013) research work is to develop a multi-objective formulation for the problem and then to adapt two solution approaches to solve the problem, i.e. ε-constraint method and the lexicographic minmax method.

4.11.2 Problem Statement

The global supply chain network of an agrichemical company consists of a production plant, several product formulation plants and a number of customer zones in different regions. This is depicted in Fig 4-17

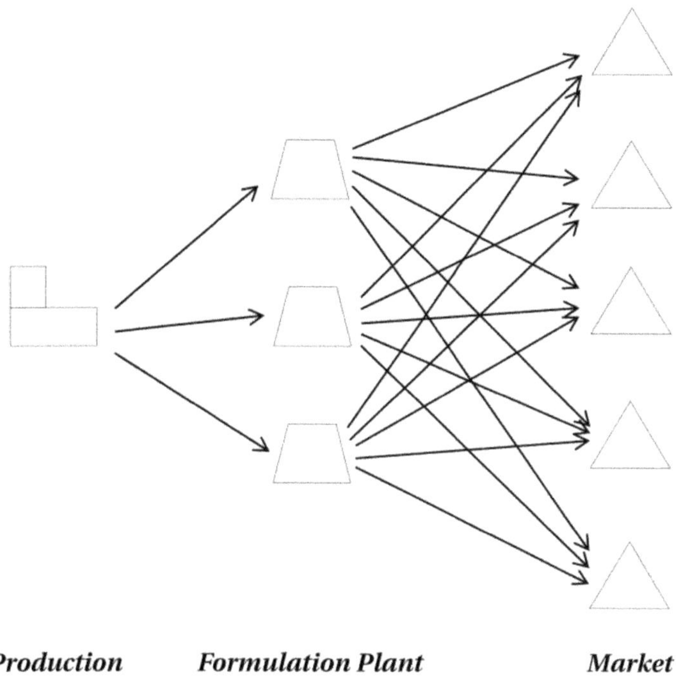

Figure 4-17: Supply chain network of an agricultural company

It is understood that the capacities of the formulation plants will not satisfy the requirements of a rapidly increasing demand, and therefore capacity extension planning is also considered. There are two optional capacity expansion strategies, i.e. proportional and cumulative capacity expansion. In the case of proportional capacity expansion planning (PCE), the maximum capacity increment of each formulation plant is proportional to the capacity prior to expansion, which means that formulation plants with larger, former capacities will have greater capacity for expansion. In the case of cumulative capacity expansion planning (CCE), capacity increment is limited to capacity prior to expansion.

Therefore the objective of this work is to determine the optimal production, distribution and capacity for the supply chain network, taking into account the cost, responsiveness and customer service levels, which may be described as follows:

1) Minimise total costs including raw material costs, formulation costs, transportation costs, inventory costs and duties

2) Minimise total response time

3) Minimise total lost sales

4.11.3 Mathematical Formulation

Following the work of Liu and Papageorgiou (2013), the nomenclature used in their model is defined in Table 4-36, and the mathematical modelling that is applied, follows thereafter:

Table 4-36: Nomenclature Table for the Multi-Objective Pareto optimisation of supply chains - Liu and Papageorgiou (2013),

Indices	Physical Meaning
g	product group
i	product
j	formulation plant
k	market
t	time period
Sets	Physical Meaning
C_j	set of groups that can be processed at formulation plant j
I_g	set of products in group g
I_k	set of products in market k
Parameters	Physical Meaning
CAP_j	capacity at formulation plant j
D_{ikt}	demand of product i in market k in period t
DC_{ijk}	unit duties for product i from formulation plant j to market k
F_{ijk}^{max}	max flow product i from form. plant j to market k in each time period
F_{ijk}^{min}	min flow product i from form. plant j to market k in each time period
FFC_{ij}	fixed formulation cost of product i at formulation plant j
FTC_{ijk}	fixed transportation cost of product i from form. plant j to market k
IC_{ij}^F	inventory cost of product i at formulation plant j
IC_{ik}^M	inventory cost of product i at market k
$IVMax_{ij}^F$	max inventory of product i at formulation plant j
$IVMax_{ik}^M$	max inventory of product i at market k
K_i	no. markets that sell product i
MC_{ij}	material cost of product i at formulation plant j
T_{ijk}	duty rate of product I from formulation plant j to market k
P_{ij}^{max}	max production of product i at formulation plant j in each time period
P_{ij}^{min}	min production of product i at formulation plant j in each time period
T	no. of time periods

VFC_{ij}	unit variable formulation cost of product i at formulation plant j
VTC_{ijk}	unit variable transp. cost of product i from form. plant j to market k
α_i	coefficient for material cost for product i in the duty function
β_I	coefficient for variable formulation cost for product i in the duty function
γ_i	coefficient for variable transport cost for product i in the duty function
σ^{max}	maximum CCE rate
σ^{min}	minimum CCE rate
θ_j^{max}	maximum PCE rate for formulation plant
θ_j^{min}	minimum PCE rate for formulation plant
ϕ_j	minimum capacity utilisation factor for formulation plant j
τ_{jk}	transportation time from formulation plant j to market k
Variables	Physical Meaning
$ECap_j$	Capacity of formulation plant j after expansion
F_{ijkt}	flow of product from form. plant j to market k in time period t
IV_{ijt}^F	inventory of product i at form. plant j in time period t
IV_{ikt}^M	inventory of product i at market k in time period t
LS_{ikt}	lost sale of product i at market k in time period t
P_{ijt}	amount product i made at form. plant j in time period t
S_{ikt}	sale of product i at market k in time period t
TDC	total duties cost
TFC	total formulations cost
TIC	total inventory cost
TMC	total raw material cost
TTC	total transportation cost
z_1	objective 1 total cost
z_2	objective 1 total flow time
z_3	objective 1 total lost sales
ΔCap_j	capacity increment for formulation plant j
Binary Variables	Physical Meaning
E_{ij}	1 if product i is produced at formulation plant j, 0 otherwise
W_{ijt}	1 if product i is assigned to formulation plant j, 0 otherwise
X_{ijk}	1 if product i is shipped from formulation plant j to market k, 0 otherwise
Y_{ijkt}	1 if product i is assigned from formulation plant j to market k, 0 otherwise

a) Production and flow constraints

The amount of product i issued to formulation plant j, during time t, is limited according to minimum and maximum production limits, i.e.

$$P_{ij}^{min}.W_{ijt} \leq P_{ijt} \leq P_{ij}^{max}.W_{ijt} \qquad \forall j, g \in G_j, i \in I_g \cap I_k, t \qquad (1)$$

If product i is shipped from formulation plant j to market k during time t, shipment vol. is limited by the minimum and maximum flow limits, otherwise the flow, $Y_{ijkt} = 0$:

$$F_{ijk}^{\min} Y_{ijkt} \leq F_{ijkt} \leq F_{ijk}^{\max} Y_{ijkt}, \qquad \forall j, k, g \in G_j, i \in I_g \cap I_k, t \qquad (2)$$

b) Inventory constraints

The inventory of product i at formulation plant j at the end of time t is equal to the inventory at the end of the previous time period plus the production and less the outgoing flows

$$IV_{ijt}^{F} = IV_{ijt-1}^{F} + P_{ijt} - \Sigma_{k;i \in Ik} F_{ijkt}, \qquad \forall j, g \in G_j, i \in I_g t \qquad (3)$$

The inventory of product i at market k at the end of t is equal to the inventory of the previous time period plus any incoming flows less sales

$$IV_{ikjt}^{M} = IV_{ikt-1}^{M} + \Sigma_{g;i \in Ig} \Sigma_{g;i \in Gj} F_{ijkt-\tau jk} - S_{ikt}, \quad \forall k, i \in I_k t \qquad (4)$$

The inventories indicated by (3) & (4) represent the initial inventory status at the formulation plants and the market respectively and are limited by upper and lower bounds accordingly:

$$IVMin_{ij}^{F} \leq IV_{ijt}^{F} < IVMax_{ij}^{F}, \qquad (5)$$

$$IVMin_{ik}^{M} \leq IV_{ikt}^{M} < IVMax_{ik}^{M}, \qquad (6)$$

c) Lost sales constraints

The sales of each product i during each time period t should not exceed corresponding demand, whilst the unsatisfied demand amounts to lost sales

$$S_{ikt} \leq D_{ikt}, \qquad \forall k, i \in I_k\, t \qquad (7)$$

$$D_{ikt} - S_{ikt} \leq LS_{ikt}, \qquad \forall k, i \in I_k\, t \qquad (8)$$

d) Proportional Capacity Expansion (PCE) constraints

The capacity increment for a plant that is expanding should be no less than the lower bound and no greater than the upper bound:

$$\theta_j^{min}\, CAP_j \leq \Delta Cap_j \leq \theta_j^{max}\, CAP_j \qquad (9)$$

e) Cumulative Capacity Expansion (CCE) constraints

Total capacity increment does not exceed total capacity, prior to expansion, multiplied by the maximum expansion rate but no less than the total capacity, prior to expansion, multiplied by the minimum expansion rate:

$$\sigma^{min} \Sigma_j CAP_j \leq \Sigma_j \Delta Cap_j \leq \sigma^{max} \Sigma_j CAP_j \qquad (10)$$

f) Capacity utilisation constraints

The capacity after expansion is equal to the capacity before expansion plus the capacity increment;

$$ECap_j = CAP_j + \Delta Cap_j, \qquad \forall j \qquad (11)$$

$$\phi_j . ECap_j \leq \Sigma_{g=Cj} \Sigma_{i=Ig} P_{ijt} \leq ECap_j \qquad (12)$$

,where ϕ_j is the minimum capacity utilisation factor between the interval (0, 1)

g) Logical constraints

If product is not assigned to the formulation plant, then it should not be able to be produced in the formulation plant:

$$\Sigma_t W_{ijt} \leq T.\ E_{ij}, \quad \forall j, g \in C_j, i \in I_g \tag{13}$$

If the transportation link for product i from formulation plant j to market k is not set up, product i cannot be shipped in any other time period

$$\Sigma_t W_{ijt} \leq T.\ X_{ijk}, \forall k, g \in C_j, i \in I_g \cap I_k \tag{14}$$

If product i is not assigned to formulation plant j, its transportation links from formulation plant j to any market k should not be established.

$$\Sigma_{ki \in Ik} X_{ijk} \leq |K_i|.\ E_{ij} \qquad \forall j, g \in C_j, i \in I_g \tag{15}$$

h) Objective

There are three aspects related to the objective in this program:

- z_1 - total cost, which includes raw material cost, formulation cost (with fixed and variable cost components) and transportation cost (with fixed, variable and duty cost components)
- z_2 – total flow time, which is equal to the summation of flows from plants to market regions multiplied by their corresponding transportation times.
- z_3 – total lost sales, which is equivalent to total unsatisfied demand

i) Total cost

Total raw material cost (*TMC*) is the unit raw material cost multiplied by the production volume:

$$TMC = \sum_t \sum_j \sum_{g\in Cj} \sum_{i\in Ig} MC_{ij} \, . \, P_{ijt} \qquad (16)$$

The fixed formulation cost is only applied if a product is allocated for formulation and the variable formulation cost is determined according to unit variable cost and volume:

$$TFC = \sum_j \sum_{g\in Cj} \sum_{i\in Ig} FFC_{ij} . X_{ijk} + \sum_t \sum_j \sum_k \sum_{g\in Cj} \sum_{i\in Ig} VTC_{ijk} . F_{ijkt} \quad (17)$$

Similarly, the transportation cost is also comprised of fixed and variable costs in that the establishment of a link involves a fixed cost and the variable cost is given by the unit transportation cost multiplied by the corresponding flows:

$$TTC = \sum_j \sum_k \sum_{g\in Cj} \sum_{i\in Ig} FTC_{ijk} . X_{ijk} + \sum_t \sum_j \sum_k \sum_{g\in Cj} \sum_{i\in Ig} VTC_{ijk} . F_{ijkt} \quad (18)$$

Duties are levied when the product is imported into the market countries and the total duty cost is the unit duty cost multiplied by the imported volume:

$$TDC = \sum_t \sum_j \sum_k \sum_{g\in Cj} \sum_{i\in Ig} DC_{ijk} . F_{ijkt} \qquad (19)$$

The unit duty cost, DC_{ijk}, is a function of raw material cost, variable production cost and variable transportation cost and rate of duty:

$$DC_{ijk} = T_{ijk} \, (\alpha_i . MC_{ij} + \beta_I VFC_{ij} + \gamma_I . VTC_{ijk})$$

$$\forall j, k, g \in C_j, i \in I_g \cap I_k \qquad (20)$$

,where α_I, β_I and γ_I are duty coefficients for raw material costs, variable production costs and variable transportation costs respectively. The inventory cost is given by the unit inventory cost multiplied by the inventory level at the end of each time period:

$$TIC = \sum_t \sum_j \sum_k \sum_{g \in Cj} \sum_{i \in Ig} IC_{ij}^F . IV_{ijt}^F + \sum_t \sum_k \sum_{i \in Ig} IC_{ik}^M \ IV_{ikt}^M \quad (21)$$

Consequently, the objective of total cost is the summation of all the costs given in (16), (17), (18), (19) and (21), i.e.:

$$z_1 = TMC + TFC + TTC + TDC + TIC \quad (22)$$

ii) Total flow time

Flow time is the flow multiplied by the transportation time and therefore total flow time across all links is as follows:

$$z_2 = \sum_t \sum_j \sum_k \sum_{g \in Cj} \sum_{i \in Ig \cap Ik} (\tau_{jk} . F_{ijkt}) \quad (23)$$

iii) Total lost sales

The objective 'lost sales' is described by the summation of the lost sales of all products at all markets over all time periods:

$$z_3 = \sum_j \sum_k \sum_{i \in Ik} LS_{ikt} \quad (24)$$

Therefore the proposed multi-objective MILP formulation incorporates (22) – (24) as the objective function and can be expressed as follows:

$$\min_{x \in Q} (z_1(x), z_2(x), z_3(x))$$

,where x is the vector of decision variables and Q is the space of feasible solutions defined by (1) – (19) and (21). Eq. (9) is only for the PCE strategy while eq. (10) is only for the CCE strategy.

4.11.4 ε-constraint method

A number of different solution techniques have been developed for the determination of Pareto optimal solutions for multi-objective MILP/MINLP (Mixed Integer Linear or Non-Linear Programs). These include the ε-constraint, weighted sum, weighted metric, goal programming and the lexicographic minimax method, of which only the ε-constraint method will be discussed in this case since it is effective and the most commonly used. With the ε-constraint Pareto-optimum method, all objective functions, apart from one, are converted into constraints with upper and lower bounds (ε) applied to each of them. Consequently, the multi-objective problem is transformed as follows:

$$\min_{x \in \Omega} f_m(\boldsymbol{x})$$

$$\text{s.t:} \quad f_p(\boldsymbol{x}) \leq \varepsilon_p, \qquad \forall p \in (1, 2 \ldots M)$$

, where $\boldsymbol{x} \in Q$ is called a Pareto optimal solution of the multiobjective problem where only the one objective function, $f_m(\boldsymbol{x})$, is minimised, while all the other objective functions are transformed into ε-constraints with their corresponding upper bounds, ε_p, where ε_p is the optimal solution of the single objective problem, $f_p(\boldsymbol{x})$. Therefore, for the multi-objective problem, min $(z_1(x), z_2(x), z_3(x),)$, each solution of the following two sub-problems is Pareto optimal:

$$\text{Min}_{x \in Q}\ z_1(x)$$

$$\text{s.t.:} \quad z_2(x) \leq \varepsilon_2 (= z_2(\boldsymbol{x}))$$

$$z_3(x) \leq \varepsilon_3\ (= z_3(\boldsymbol{x}))$$

, and

$$\text{Min}_{x \in Q}\ z_1(x)$$

$$\text{s.t.:} \quad z_2(x) \leq \varepsilon_2 (= z_2(\boldsymbol{x}))$$

$$z_3(x) \leq \varepsilon_3\ (= z_3(\boldsymbol{x}))$$

4.11.5 Extraction of the optimisation technique for a bi-objective supply chain under uncertainty

Table 4-37: Optimisation technique for a bi-objective supply chain under fuzzy uncertainty based on the work of Liu and Papageorgiou (2013)

1	Description and illustration of supply chain network
2	Requirements Definition: Minimise production costs, supply chain flow times and lost sales
3	Nomenclature Requirements
4	Define type of Uncertainty involved and Solution approach: Fuzzy uncertainty is associated with the supply chain (production and distribution) aspects, whilst multi-objective uncertainty is associated with the need to minimise the costs, supply chain flow times and lost sales in a Pareto optimal fashion. There is also a need to consider and incorporate capacity utilisation and expansion constraints
4	Formulate the multi-objective fuzzy mixed integer linear program - Assign the principal objective - Allocate the remaining objectives as Pareto ε-constraints - Specify constraints
5	Solve the MILP
6	Record (Pareto) optimum results

4.11.6 Concluding Remarks

This paper by Liu and Papageorgiou (2013) discussed the optimisation of a supply chain problem in which the production, distribution and capacity expansion decisions were optimised while considering total cost, total flow time and total lost sales as objectives. The ε-constraint Pareto-optimum method was applied to solve the multi-objective problem in which the 'total cost' function was assigned as the single objective to be optimised while the total flow and total lost sales functions were transformed into constraints.

The supply chain, under uncertainty, optimisation technique utilised herein was extracted and is recorded in Table 4-37

4.12 A bi-objective stochastic fuzzy model for a warehouse in a supply chain network

- based on the work of Razmi et al (2013) -

4.12.1 Introduction

In a paper by Razmi et al. (2013), the fact was recognised that the traditional supply chain network design approach was changing in reflection of the increasing trend towards globalisation. As opposed to the traditional network design approach, any current redesign approach also should take into account any prevailing conditions or constraints that might impact the design. Chief amongst these is the design of warehouses in terms of their number, respective capacities and reliabilities and customer service levels.

Literature research had also revealed that the following shortcomings with regard to warehousing:

- The importance of warehouse reliability was ignored
- Zone specific warehouse criteria, which include environmental or capacity restrictions, were ignored.
- Only one type of delivery lead time (soft) had been taken into account but the other (hard), taking into account service disruptions, had to be considered.

The purpose of this paper, by Razmi et al. (2013), is to address these shortcomings in an integrated model. The cost trade-off with two different delivery times, in conjunction with uncertain demand, is covered by means of a novel bi-objective stochastic optimisation solution approach.

4.12.2 Formulation of Model

Further to the work of Razmi et al. (2013), a representative two echelon supply chain network consisting of plants, distribution centres and customer zones, as depicted in Fig 4-18 was used as the basis for their research.

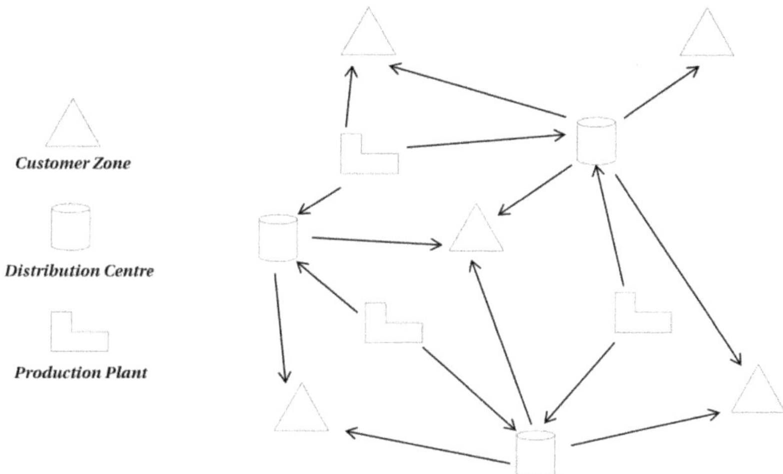

Figure 4-18: Two echelon Supply chain network

Two delivery lead times exist in this network, the first being the expected lead time, during which the demand of all customers should be met, and the second being the preferred delivery time of customers. The objective of this exercise is the capacity consolidation of existing warehouses plus the identification of new sites for capacity transfer or capacity generation in order to determine the optimal flow at minimum cost. Moreover, the maximisation of customer demand is sought after. Consequently, a bi-objective problem arises, i.e. minimisation of network cost plus maximisation of customer demand.

Supply Chain Network Assumptions

- Demand, transportation and warehousing costs are random
- Product shortfall is not permitted in the network
- Both the available and potential warehouses have deterministic fixed costs, which, in the case of existing facilities, equate to set-up costs and which, in the case of new facilities, equate to installation costs.
- When a warehouse is consolidated into another warehouse, its entire capacity is relocated.

Table 4-38: Nomenclature for stochastic warehouse modelling - Razmi et al. (2013)

Indices	Physical Meaning
i	fixed locations of production zones, $i \in I$
j	set of warehouse locations, $j \in J$
j'	fixed locations of warehouse zones, $j' \in J'$
j^*	set of potential warehouse locations, $j^* \in J^*$
k	fixed locations of customer zones
Parameters	**Physical Meaning**
d_{ks}	demand of customer k for scenario s
fvw_j	cost per unit capacity of warehouse j
ffw_j	fixed cost of maintaining warehouse j exc. capacity cost
rw_{j^*j}	moving and relocating cost of unit capacity j^* to consolidating site j
sfw_j	amount saved from closing existing warehouse j
tc_{ij}^s	unit transport cost between plant i and warehouse j for scenario s
tc_{jk}^s	unit transport cost between warehouse j and cust. k for scenario s
cwf_j	throughput capacity of existing warehouse j
cpf_i	production capacity of plant i
caw_j	max warehouse capacity accepted at site j
t_{jk}	transhipment time from warehouse j to customer k
τe	max expected delivery lead time from warehouses
p_s	probability of scenario s
Variables	**Physical Meaning**
Z_{j^*j}	'1', if available warehouse capacity j^* is relocated to warehouse j or '0' if available warehouse capacity is not affected
a_j^s	'1', if warehouse j is inaccessible, otherwise '0' for scenario s
od_{jk}	'1' if access time of warehouse j to cust. k is more than, otherwise '0'
W_{j^*}	'1' if a hybrid warehouse-collection centre is opened at location j^*, otherwise '0'.

$$\min Z_1 = \sum_{s \in S} p_s \, Cs^s \quad (1)$$

$$\max Z_2 = \sum_{s \in S} p_s \, (d_{ks} - od_{jk} \sum_{j \in J} X_{jk}^s) / \sum_{k \in K} d_{ks} \quad (2)$$

, subject to:

$$Cs^s = \sum_{j^* J^*} ffw_{j^*} W_{j^*} + \sum_{j^* \in J^*} \sum_{j \in J} rw_{j^*j} Z_{j^*j} + \sum_{j^* \in J^*} ffw_j Z_{j^*j} + \sum_{j^* \in J^*} fvw_j \sum_{j^* \in J^*} cwf_j Z_{j^*j} - \sum_{j^* \in J^*} ffw_{j^*} \sum_{j \in J} Z_{j^*j} - \sum_{j^* \in J^*} sfw_j (1 - \sum_{j \in J} Z_{j^*j}) + \sum_{i \in I} \sum_{j \in J} tc_{jk}^s X_{jk}^s, \quad \forall s \in S \quad (3)$$

$$\sum_{j \in J} Y_{ij}^s \leq cpf_i, \quad \forall s \in S,\ i \in I \quad (4)$$

$$\sum_{i \in I} Y_{ij}^s = \sum_{k \in K} X_{jk}^s, \quad \forall s \in S,\ j \in J \quad (5)$$

$$\sum_{j \in J} X_{jk}^s = d_{ks} \quad \forall s \in S,\ k \in K \quad (6)$$

$$\sum_{j^* \in J^*} Z_{j^*j} \leq |J^*|\, Z_{j^*j}, \quad j \in J \quad (7)$$

$$\sum_{j^* \in J^*} Z_{j^*j} \leq |J^*|\, W_{j^*}, \quad j^* \in J^* \quad (8)$$

$$\sum_{j \in J} Z_{j^*j} \leq 1, \quad j^* \in J^* \quad (9)$$

$$\sum_{k \in K} X_{jk}^s \leq (1 - a_j^s) \sum_{j^* \in J^*} cwf_j Z_{j^*j}, \quad \forall s \in S,\ j \in J \quad (10)$$

$$\sum_{j^* \in J^*} cwf_j Z_{j^*j} \leq caw_j \quad (11)$$

$$\sum_{s \in S} \sum_{j \in J} \sum_{k \in K} od_{jk} X_{jk}^s = 0 \quad (12)$$

$$X_{jk}^s,\ Y_{ij}^s \geq 0 \quad (13)$$

$$Z_{j^*j},\ W_{j^*} \in (0\ 1)$$

There are two objectives with this model. The first one involves the minimisation of the total costs across all scenarios, and cognisance is taken of the probability of occurrence each scenario. The second objective involves the maximisation of customer demand coverage over the area concerned.

Constraint (3) indicates the network costs per scenario, constraint (4) signifies the production capacity of each plant while constraint (5) reflects the fact that the input flow to each warehouse must equal the output flow each warehouse. Constraint (6) represents customer demand satisfaction while constraint (7) ensures that, in the event of the capacity of one warehouse, e.g. (*a*), consolidating with that of another warehouse, e.g. (*b*), the capacity of (*b*) is then not allowed to reconsolidate back into (*a*) again. Constraint (8) indicates that if a new warehouse is established, the total capacity it can receive from other warehouses must not exceed its own capacity levels. Constraint (9) states that the capacity of no more than one warehouse may be transferred at a time while constraint (10) indicates that the product transferred from one warehouse may not exceed its total capacity. Constraint (11) represents the condition that the total capacity consolidated into any one warehouse must not exceed the zone capacity; constraint (12) represents the condition that shipment duration for any one customer may not be greater than the expected delivery time while constraint (13) enforces continuous and discrete restrictions on associated decision variables.

4.12.3 Solution Procedure

Since there are two, potentially conflicting, objectives involved in this problem, i.e. minimisation of costs and maximisation of customer demand satisfaction, a Pareto-optimal approach, wherein any one objective cannot be improved without detrimentally affecting another, is adopted for this problem. On account of certain solution inefficiencies with the traditional ε-constraint Pareto-optimal solution approach, especially when more than two objectives are involved, an augmented ε-constraint approach was put forward.

Such augmented ε-constraint approach involved the consolidation and insertion of slack and surplus variables, from the constraint functions, as a new term in the objective function and, in so doing, removing the inefficiencies that were previously incurred. The problem of longer solution time, which was incurred in the case of more than two objective functions, was overcome by the construction of a payoff table. This table consists of the 'best' and the 'worst' values for each objective function and, consequently, the augmented ε-constraint approach can be described as follows:

$$\text{Max}\, f_1(p) + eps \sum_{j=1} (s_j / r_j)$$

subject to:

$$p \in F$$

$$f_1(p) - s_j = \varepsilon_j, \forall j \neq i$$

$$s_j \in R^+$$

, where:

- F is the feasible space of the original problem
- $f_1(p)$ is the vector of objective functions
- s_j is the slack variables of the j-th constrained objective function
- r_j is the range of constrained objective functions
- *eps* is a very small number, between 10^{-6} and 10^{-3}

1) The range of each objective function will be used as constraints, derived from the payoff table

2) The range of each objective function is divided into m equal intervals and hence $(m_j + 1)$ grid points will emerge. These points are used to vary the right-hand side of the corresponding objective functions

3) The augmented ε-constraint model is solved for each vector of ε, which is obtained from step 2)

4.12.4 Extraction of the bi-objective stochastic optimisation technique for a supply chain under uncertainty

Table 4-39: Extraction of the bi-objective stochastic optimisation technique for a Supply Chain under uncertainty based on the work of Razmi et al. (2013)

1 Description of the supply chain network including production, warehousing, distribution and including the type and nature of two delivery times

2 Objective of the problem – Consolidation of warehouse capacity in order to minimise total cost, whilst maximising delivered demand satisfaction

3 Specification of all parameters and variables involved in the modelling of the supply chain network

4 List all specifications and assumptions

5 Demand uncertainty is represented by a number of different probabilistic scenarios

6 Formulation of the bi-objective stochastic program MINLP model
- Define and specify both objective functions
- Define and specify all constraints

7 Follow the Pareto ε-constraint method for a bi-objective MINLP, as put forward by Haimes and Chankong (1979), so that both objectives, between them, achieve an optimum solution
- Select the principal objective as the single MINLP objective function
- The other, secondary objective, is assigned as the ε-constraint
- Solve the model

8 Record the optimum results

4.12.5 Conclusion

This paper addressed the formulation and solution of a two-stage bi-objective stochastic mixed integer linear programming model, with regard to simultaneously minimising operating costs and maximising customer demand levels, from the perspective of redesigning the warehouse chain in terms of consolidating capacities and minimising service disruption levels. The bi-objective stochastic optimisation technique utilised was extracted and tabulated in Table 4-39.

4.13 The Optimisation of a Supply Chain Network (SCN) under Stochastic Fuzzy uncertainty

- based on the work of Awudu and Zhang (2013) -

4.13.1 Introduction

As a result of increasingly high worldwide energy consumption, renewable energy sources, such as biofuels, have been considered as alternative energy sources, and especially in the transportation sector. However, in order to ensure a regular and consistent supply of biofuels to the market, a reliable and resilient supply chain is required. Also, since previous research works, in following a deterministic approach, did not adequately cover supply chain uncertainties, this paper will consider a combined deterministic/stochastic approach. The objective of this paper is therefore to maximise the operating profit of a supply chain network, consisting of production plants and distribution centres, subject to uncertain prices and demand. The problem is modelled as a stochastic programming problem in two stages, with the first stage representing the production aspects and the second stage representing the distribution aspects. The Benders decomposition with Monte Carlo simulation is used to solve the proposed model. The prices of end products follow the Geometric Brownian Motion (CBM)

4.13.2 Problem Description

The biofuel supply chain network discussed in this case is shown in Fig 4-19. It consists of biomass raw material sources, (*k*) biofuel refinery plants and (*c*) distribution centres

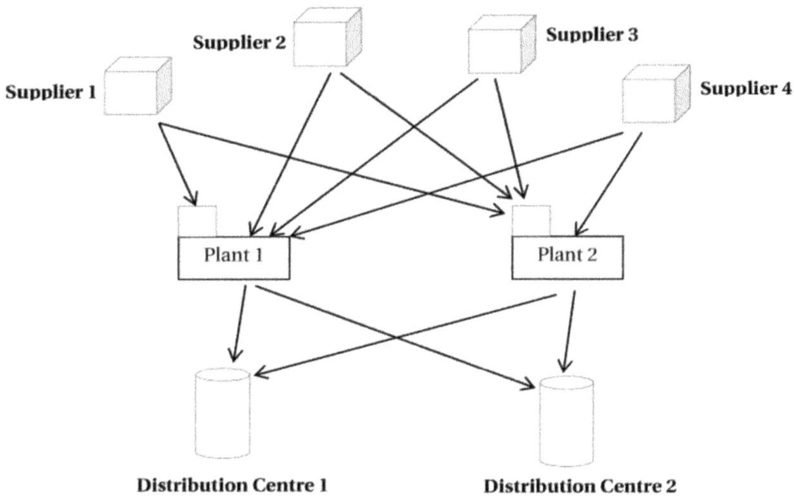

Figure 4-19: Biofuel Supply Chain

The problem objective is to maximise plant profit by taking into account raw material purchases, raw material usage rate, production rate, product sold and raw material and final product inventory levels. A stochastic linear programming model is proposed whereby uncertainties are incorporated in market demand and in the price of finished goods. Nomenclature utilised is defined in Table 4-40

Table 4-40: Nomenclature for stochastic fuzzy modelling – Awudu and Zhang (2013)

Index	
i	represents supply sources of raw materials, $i = 1, 2..I$
j	represents the product, $j = 1, 2...J$
k	represents the refinery, $k = 1, 2....K$
c	represents the customer distribution centres, $c = 1, 2...C$
Decision variables	
$x_{i,k}$	amount of biomass raw materials from supply source i to plant k
$s_{j,k}$	amount of product j sold at plant k
$z_{j,k}$	amount of product j produced at plant k
$v_{i,k}$	amount of raw material from supply source i consumed at plant k
$R_{i,j}$	amount of raw material inventory from supply source i for product j
$F_{j,k}$	amount of end products j at plant k
$L_{j,k}$	amount of lost sales j of product j at plant k
$B_{j,k}$	amount of backlog of product j at plant k
Parameters	
$P_{j,k}$	selling price of product j at plant k
$\theta_{I,k}$	available biomass from the supply source i at plant k
$\Delta_{j,k}$	conversion factor for the end product j at plant k
$y_{j,k}$	unit cost of raw materials from supply source i at plant k
$\tau_{i,k}$	unit transportation cost of raw materials from supply source i to plant k
$\tau_{k,c}$	unit transportation cost of end products from plant k to demand point c
$d_{i,k}$	transportation distance of raw materials from supply source i to plant k
$d_{k,c}$	transportation distance of end products from plant k to demand point c
$F^{0}_{j,k}$	end products inventory for previous period at plant k for end product j
$R^{0}_{j,k}$	raw material inventory for previous period at plant k for RM from source i
$B^{0}_{j,k}$	amount of backlog for previous period at plant k for product j
$L^{0}_{j,k}$	amount of lost sale for previous period at plant k for product j
β_{i}	cost of processing raw material from supply source i
$m_{j,k}$	cost for lost sales for product j at plant k
$q_{j,k}$	cost for backlog for product j at plant k
$D_{j,k}$	demand for product j at plant k

$\upsilon_{i,k}$	amount of raw materials consumed from supply source *i* at plant *k*
$\mathfrak{J}_{j,k}$	amount of fractional loss in demand for product *j* at plant *k*
S_o	Initial spot price of the end products
Stochastic variables and terms	
ε	Scenario representation for the stochastic variable
ρ_ε	probability of the scenario of each stochastic variable
$S_{jk\varepsilon}$	stochastic sales amount of product j at plant k based on scenario ε
$B_{jk\varepsilon}$	Stochastic backlog for previous period at plant k for product j based on scenario ε
$F_{jk\varepsilon}$	stochastic end products inventory in previous period at plant *k* for product *j* for scenario ε
$L_{jk\varepsilon}$	stochastic lost sale for previous period at plant *k* for product *j* for scenario ε
$P_{jk\varepsilon}$	stochastic price of finished goods of product *j* at plant *k* for scenario ε
S_c	calculated spot price of the end products after the price scenarios have been generated
I	iteration steps for Benders decomposition
N	total number of scenarios generated

4.13.3 Deterministic Model

Following the work of Awudu and Zhang (2013), in the deterministic model, since product demand and pricing are constants, the mean or average values thereof are used. Consequently, the objective function, which maximises production profit by takes into account raw material acquisition, raw material consumption, production costs and raw material and finished product inventory levels. The total costs, which include the raw material purchasing costs, the transportation costs and the inventory holding costs, are expressed in the form of constraints. The entire MILP formulation which, according to Awudu and Zhang (2013), is expressed below as follows:

Objective Function:

Max $[\sum^{J}_{j=1}\sum^{I}_{k=1}P_{j,k}S_{j,k} - \sum^{I}_{i=1}\sum^{K}_{k=1}y_{i,k}x_{i,k} - \sum^{I}_{i=1}\sum^{K}_{k=1}\tau_{ik}d_{i,k}x_{i,k} - \sum^{K}_{k=1}\sum^{C}_{c=1}$ $\sum^{J}_{j=1}\tau_{kc}d_{kc}z_{j,k} - \sum^{I}_{i=1}\sum^{K}_{k=1}h_{i,k}R_{i,k} - \sum^{J}_{j=1}\sum^{K}_{k=1}h_{j,k}F_{j,k} - \sum^{I}_{i=1}\sum^{K}_{k=1}\beta_{i,k}v_{i,k} -$ $\sum^{J}_{j=1}\sum^{K}_{k=1}m_{j,k}L_{j,k} - \sum^{J}_{j=1}\sum^{K}_{k=1}q_{j,k}B_{j,k}]$

subject to Constraints:

i) Capacity constraints:

$x_{i,k} \leq \theta_{ik}, \ \forall i, \forall k$

$v_{i,k} \leq x_{ik}, \ \forall i, \forall k$

ii) Demand constraints

$B_{j,k} + D_{j,k} = s_{j,k} + L_{j,k} + B_{j,k}, \qquad \forall j, \forall k$

$L_{j,k} + \Im_{j,k}(D_{j,k} + B_{j,k} - s_{j,k}) \qquad \forall j, \forall k$

iii) Raw material and finished products inventory constraints

$R_{i,k} = R^{0}_{i,k} + x_{i,k} - v_{i,k} \qquad \forall j, \forall k$

$F_{j,k} = F^{0}_{j,k} + z_{j,k} - s_{j,k} \qquad \forall j, \forall k$

iv) Material balance constraints

$z_{j,k} = \Delta_{i,j,k}(\sum v_{i,j,k}), \qquad \forall i, \forall j, \forall k$

v) Sales, demand and production balance constraints

$s_{j,k} \leq z_{j,k} \qquad \forall j, \forall k$

$s_{j,k} \leq D_{j,k} \qquad \forall j, \forall k$

$x_{i,k}, z_{j,k}, D_{j,k}, v_{i,k}, s_{j,k}, R_{i,k}, F_{j,k}, \theta_{ik}, L_{j,k}, B_{j,k} \geq 0, \qquad \forall i, \forall j, \forall k$

4.13.4 Stochastic Model

A stochastic linear program, which amounts to a stochastic enhancement to the prior deterministic version, is put forward to account for uncertainties in decision making, which, in this case, amount to probabilistic scenarios. Therefore:

Objective Function:

Max $[\Sigma^{J}_{j=1}\Sigma^{N}_{\varepsilon}\ \Sigma^{I}_{k=1}pP_{j,k}S_{j,k}\ -\ \Sigma^{I}_{i=1}\Sigma^{K}_{k=1}y_{i,k}x_{i,k}\ -\ \Sigma^{I}_{i=1}\Sigma^{K}_{k=1}\tau_{ik}d_{i,k}x_{i,k}\ -\ \Sigma^{K}_{k=1}\Sigma^{C}_{c=1}\ \Sigma^{J}_{j=1}\tau_{kc}d_{kc}z_{j,k}\ -\ \Sigma^{I}_{i=1}\Sigma^{K}_{k=1}h_{i,k}R_{i,k}\ -\ \Sigma^{J}_{j=1}\Sigma^{N}_{\varepsilon}\ \Sigma^{K}_{k=1}phj_{,k}F_{j,k}\ -\ \Sigma^{I}_{i=1}\Sigma^{K}_{k=1}\beta_{i,k}v_{i,k} - \Sigma^{J}_{j=1}\Sigma_{\varepsilon}^{N}\ \Sigma^{K}_{k=1}pmj_{,k}L_{j,k}\ - \Sigma^{J}_{j=1}\Sigma^{N}_{\varepsilon}\ \Sigma^{K}_{k=1}pqj_{,k}B_{j,k}]$

subject to Constraints:

i) Capacity constraints:

$x_{i,k} \leq \theta_{ik},\ \forall i, \forall k$

$v_{i,k} \leq x_{ik},\ \forall i, \forall k$

ii) Demand constraints

$B_{j,k} + D_{j,k} = s_{j,k} + L_{j,k} + B_{j,k},\qquad \forall j, \forall k$

$L_{j,k} + \Im_{j,k}(D_{j,k} + B_{j,k} - s_{j,k})\qquad \forall j, \forall k$

iii) Raw material and finished products inventory constraints

$R_{i,k} = R^{0}{}_{i,k} + x_{i,k} - v_{i,k}\qquad \forall j, \forall k$

$F_{j,k} = F^{0}{}_{j,k} + z_{j,k} - s_{j,k}\qquad \forall j, \forall k$

iv) Material balance constraints

$z_{j,k} = \Delta_{i,j,k}(\Sigma v_{i,j,k}),\qquad \forall i, \forall j, \forall k$

v) Sales, demand and production balance constraints

$s_{j,k} \leq z_{j,k}\qquad \forall j, \forall k$

$s_{j,k} \leq D_{j,k}\qquad \forall j, \forall k$

$x_{i,k},\ z_{j,k},\ D_{j,k},\ v_{i,k},\ s_{j,k},\ R_{i,k},\ F_{j,k},\ \theta_{ik},\ L_{j,k},\ B_{j,k} \geq 0,\qquad \forall i, \forall j, \forall k$

Uncertainties in price and demand

In this particular case, normal distributions are used to model uncertain demands and Geometric Brownian Motion (GBM) is used to model uncertain prices. The basic Brownian motion model equation is as follows:

$$dS_t/S_t = \mu dt + \sigma dz \qquad (1)$$

, where the commodity price at time t is represented by a mean value, (μ), and volatility, (σ). The spot price at time is expressed as S_t. The term z is a Wiener process, where $dz = \varepsilon\sqrt{dt}$ and ε is the uncertain term. The term, dt, represents the change in time. All terms are derivatives, and not partial derivatives. Scenarios are generated by using the derived form of (1), i.e.:

$$S_t = S_o(1 + \mu dt + \varepsilon\sigma\sqrt{dt})$$

4.13.5 Proposed Solution: Benders decomposition with Monte Carlo simulation

Since stochastic programming models, which are based on a number of different operating scenarios, are frequently very computationally difficult to solve, an additional algorithm was introduced to minimise such computational complications, and such algorithm is called Benders decomposition. By this technique, problems are broken down into master and sub-problems, where the master contains the deterministic component while the sub-problem contains the stochastic portion. The solution of the sub-problem requires that the Monte Carlo simulation technique is applied, which is achieved as follows, below:

Sub-problem:

$$MinSub = - \sum^{J}_{j=1}\sum^{N_\varepsilon} \sum^{I}_{k=1} p P_{j,k} S_{j,k} + \sum^{J}_{j=1}\sum^{N_\varepsilon} \sum^{K}_{k=1} p h_{j,k} F_{j,k} + \sum^{J}_{j=1}\sum_{\varepsilon}^{N} \sum^{K}_{k=1} p m_{j,k} L_{j,k} + \sum^{J}_{j=1}\sum^{N_\varepsilon} \sum^{K}_{k=1} p q_{j,k} B_{j,k}$$

, subject to constraints

$D_{j,k}$ - $B_{j,k}$ = $s_{j,k}$ + $L_{j,k}$ - $B_{j,k}$, $\forall j, \forall k$

$\Im_{j,k}\, s_{j,k}$ - $\Im_{j,k}\, B_{j,k}$ = $\Im_{j,k} D_{j,k}$ – $L^{o}_{j,k}$ $\forall j, \forall k$

$F_{j,k}$ + $z_{j,k}$ - $s_{j,k}$ = $F^{0}_{j,k}$ $\forall j, \forall k$

$z_{j,k}$ - $s_{j,k}$ – slk1 = 0, $\forall j, \forall$

$s_{j,k}$ + slk2 = $D_{j,k}$ $\forall j, \forall k$

$x_{i,k}$, $z_{j,k}$, $D_{j,k}$, $\upsilon_{i,k}$, $s_{j,k}$, $R_{i,k}$, $F_{j,k}$, θ_{ik}, $L_{j,k}$, $B_{j,k} \geq 0$, $\forall i, \forall j, \forall k$

Master problem

MinOpt = $\sum^{I}_{i=1}\sum^{K}_{k=1} y_{i,k} x_{i,k}$ + $\sum^{I}_{i=1}\sum^{K}_{k=1} \tau_{ik} d_{i,k} x_{i,k}$ + $\sum^{C}_{c=1}\sum^{K}_{k=1} \tau_{ik} d_{i,k} z_{i,k}$ + $\sum^{I}_{i=1}\sum^{K}_{k=1} h_{i,k} R_{i,k}$ - $\sum^{I}_{i=1}\sum^{K}_{k=1} \beta_{i,k} \upsilon_{i,k}$ + η

, subject to constraints

$x_{i,k} \leq \theta_{ik}$, $\forall i, \forall k$

$\upsilon_{i,k} \leq x_{ik}$, $\forall i, \forall k$

$R_{i,k} = R^{0}_{i,k} + x_{i,k} - \upsilon_{i,k}$ $\forall j, \forall k$

$z_{j,k} = \Delta_{i,j,k}(\sum \upsilon_{i,j,k})$, $\forall i, \forall j, \forall k$

$\eta \geq (1/N)\sum^{N}_{\varepsilon=1} \pi_{\varepsilon}(h_{\varepsilon} - z_{j,k} B_{j,k})$ – optimality cut (2)

4.13.6 Extraction of the combined Benders decomposition and Monte Carlo simulation routine thus creating a 2 stage stochastic fuzzy optimisation technique

Table 4-41: The Two Stage Stochastic Fuzzy Optimisation Technique based on the work of Awudu and Zhang (2013)

Step	Description
1.	Set I (iteration counter) = 1, solve the Master problem, and let $\eta = 0$ in order to obtain the optimal decision values for the Master problem without cut. Also set the solution upper bound (UB) to infinity (∞) and set the solution lower bound (LB) = 0
2.	Use the Monte Carlo technique to generate N scenarios for the demand and price data by using the normal distribution and GBM respectively for all final (end) products.
3.	Solve the sub-problem by using $z_{j,k}$ as a constant to determine the optimal variables, $s_{j,k}$, $F_{j,k}$, $L_{j,k}$ and $B_{j,k}$.
4.	Determine the various scenario solutions of the sub-problem and represent them as π_{ε} scenario solutions.
5.	Update the upper bound setting by assigning $UB_{I+1} = \min[UB_I, Opt + (Sub - Opt)]$
6.	Update the lower bound by applying $LB_{I+1} = Opt$
7.	Add equation (2) to Master problem
8	Check whether ($UB_I - LB_I$) is tolerable, If so, the solution is optimal, if not, proceed with the iterations
9.	Finally solve the updated Master problem, add the updated cut, η

4.13.7 Conclusion

The research paper by Awudu and Zhang (2013) demonstrated that the optimal production planning of a multi-product biofuel plant could be achieved via applying an iterative two stage, master and sub-problem, stochastic linear programming procedure until tolerable optimum results were achieved. This technique was primarily designed for the biofuel industry. When this technique was applied to an actual biofuel supply chain in the United States, it was demonstrated that the optimum operating results achieved were better than the deterministic equivalents.

The supply chain, under uncertainty, optimisation technique utilised in this procedure is documented in Table 4-41.

4.14 Gradient algorithm for chance constrained fuzzy non-linear goal programming (incl. transformation of probabilistic functions into deterministic equivalents)

- based on the work of Lee and Olson (1985) -

4.14.1 Introduction

The taking of management decisions often involves multiple, often conflicting, objectives against backgrounds of operational uncertainty. Although, sometimes, such uncertainty may be inconsequential in management decision-making, statistically determined measures of risk are often applied when the probability distribution of outcomes can be described. The application of more precise measures of uncertainty allows for more precise predictions of decision outcome. Goal programming has been developed as a means of analysis for a wide variety of decision problems involving multiple, potentially conflicting objectives.

The consideration of risk in certain environments may lead to the generation of non-linear programming models. However, since goal programming models have been shown to be equivalent to a series of single objective models, chance constrained, or probabilistic, goal programming models could be solved by a series of conventional mathematical programming models using general nonlinear techniques such as the generalised reduced gradient codes, which are commercially available. While commercially developed codes may be effective for handling particular sub-problems, they would not necessarily incorporate the consideration of risk. Therefore separable nonlinear programming, which requires a large number of constraints to reflect nonlinear functions, was developed. However, since this technique required much input data development and also involved a lot of computational time, the goal programming algorithm was proposed by Lee and Olson (1985), which provides application flexibility for decision-makers through the use of optimal step length calculations.

4.14.2 Chance constrained goal programming

The probabilistic nature of chance constrained programming is especially useful for goal programming analyses. Goal programming enables multiple levels of confidence to be proposed for each constraint within a model and the pre-emptive ranking of objectives, by importance, provides a framework for decision makers to analyse trade-offs between alternative objectives, as well as to change confidence levels. The use of chance constraints requires the development of deterministic equivalents for corresponding probability functions. Deterministic equivalents were developed for models and were most generally used in practice. A number of illustrative uses of probabilistic modelling, including comparison of chance constraints with other methods, were provided

The general linear goal programming problem is stated as follows:

Min $Z = P_1(\Sigma^m_{i=1} w_i d_i^{+/-});\ldots; P_k(\Sigma^m_{i=1} w_i d_i^{+/-})$

subject to: $\Sigma^m_{j=1} a_{ij} x_j + d^-_i - d^+_i = b_i$, $\quad i = 1 \ldots\ldots m$,

$b_i, d^-_i, d^+_i \geq 0$, $\quad i = 1 \ldots\ldots m$,

$x_j > 0$, $\quad j = I \ldots\ldots n$

, *where* x_j - *the* j^{th} *variable*

b_i - the target for goal i

a_{ij} - the coefficient of x_j in the ith goal constraint

d^-_i - the under-achievement of goal i

d^+_i - the over-achievement of goal i

P_k - the kth pre-emptive priority level

w_i - the weight assigned to d^+_i or d^-_I at priority k

Introduction of chance constraints in goal programming yields deterministic equivalents that are non-linear when a_{ij} are stochastic. Chance constraints in the goal programming model are of the form:

$$\Pr\left\{\Sigma^{n}_{j=1} a_{ij} x_j \leq b_i\right\} \geq \alpha_I, \qquad 0 < \alpha_I < 1$$

As parameters a_{ij} and b_j may both be random variables, the chance constraint can be converted to its deterministic equivalent as:

$$E\left\{\Sigma^{n}_{j=1} a_{1j} x_j\right\} + F^{-1}(\beta)\left\{V\Sigma^{n}_{j=1} a_{1j} x_j - b_i\right\}^{1/2} \leq E(b_i)$$

, where E( ) and V( ) are expected value and variance operators respectively. F( ) is the cumulative standardised normal probability distribution function of:

$$\left\{\frac{(\Sigma^{n}_{j=1} a_{1j} x_j - b_i) - E(\Sigma^{n}_{j=1} a_{1j} x_j - b_i)}{V(\Sigma^{n}_{j=1} a_{1j} x_j - b_i)}\right\}^{1/2}$$

, and F^{-1} is the inverse of $F(\,)$, e.g. $F^{-1}(0.95) = 1.645$

Chance constraints in goal programming formulations yield deterministic equivalents that are non-linear. The objective is, therefore, to identify feasible points for goal constraints that are subject to structural constraints. Consequently, the concept of feasible directions can aid in the evaluation of non-linear constraints within goal programming formulations because this method requires that the constraints form a convex solution space at each goal level. Since chance constraints involving normally distributed variance elements are monotonic in nature, the requirement for convexity is satisfied and therefore separable programming methods can be applied to chance constrained goal programming. However, if that approach suffers from the need to modify the model to include breakpoints, the accuracy of that method can be limited by the size of grid selected. The algorithm developed by such study also requires significant computer time and space. The gradient method is another nonlinear approach suitable for chance constrained models. While the generalized reduced gradient method would provide a more efficient solution method for each step of a series of single objective models, it would involve, potentially, a large number of steps. The goal programming model, however, could be solved in one step by the proposed algorithm, below:

4.14.3 The gradient algorithm for non-linear goal programing

The proposed gradient algorithm utilises the gradient of constraints to identify and initiate a feasible direction. A step length routine enables the optimal distance to be calculated, and then that such movement can occur. Such step length routine involves the consideration of feasible bounds of structural constraints first before, dealing with goal constraints, in order of priority. Should a goal constraint be encountered that cannot be satisfied within the bounds of prior constraints, then step size, *r*, is determined by a minimum deviation from the goal target. This new point is compared with the solution point obtained in the previous iteration, and if the difference is less than a selected level, the algorithm terminates. However, if the reverse applies, the gradient of all constraints are evaluated at the new point, and the procedure continues. The algorithm contains five steps, and is shown in Figure 4-15. The starting point for the algorithm is the origin. At this point, the gradient of non-linear constraints is equal to the expected value of the technological coefficients.

Gradient Algorithm

Step 1: *Determine normalized constraints.* For subsequent iterations of the algorithm, the gradient vectors of nonlinear constraints are substituted for the technological coefficients to identify the half spaces that yield improved nonlinear function values. In order to avoid the potential problem of zigzagging a zero-one parameter, θ is created for all non-linear constraints. If a non-linear constraint is binding, θ is set equal to one for that constraint. This perturbs the normalized gradient of that constraint a small distance, avoiding algorithm convergence on a suboptimal point.

Step 2: *Determine an improving direction.* The modified simplex method of goal programming is used to identify a direction of improvement by solving the constraint set identified in step 1 according to the priority structure of the model. The goal programming solved it:

$$\text{Min Z} = P_1\left\{\Sigma_{i=1}{}^{m} wd_i^{+/-}\right\}; \ldots; P_k\left\{\Sigma_{i=1}{}^{m} wd_i^{+/-}\right\}$$

, where $P_k. \gg P_{k-1}$.for any *k*. and w_i are weights within priority levels,

s.t. $(q_p + \theta a_{1j})'x_j + d_i^- - d_i^+ = b_i + \theta a_{1j}x_j$, x_j, d_i^-, $d_i^+ \geq 1$ o,

, and q_p is the gradient vector for constraint *i*, *x* is the solution vector. Solution of this goal programming model identifies a new solution point X_p^*

Step 3: *Determine optimal direction.* The calculation G = X~- Xp_~ determines the direction for this iteration.

Step 4: *Determine optimal step length.* The optimal step length is determined by linear search in the direction obtained from Step 3 according to the priority structure of goals. Upper and lower bounds for step lengths for each of these constraints are determined and used to restrict the step length searched for lower goals. If a goal constraint is found that is not satisfied over the bounded step length, the step length resulting in a minimum deviation from that goal is selected as the optimal step length. A maximum upper bound of 1 is included in the algorithm to guarantee feasibility at each iteration. The optimal step length S e is used in calculating the next solution point:

$$X_p = S_p(U_p) + X_{p-1}$$

Should all constraints be satisfied by X_p, the solution for the problem is identified by X_p, There will possibly be more than one point satisfying all constraints in this case, but all of these points would provide equal satisfaction of goals. Such an occurrence would indicate that goal levels were not as high as they could have been,

Step 5: If the new solution X_p, is within a selected parameter ε (set to 0.001 in the algorithm) of X_{p-1} for all x_i, the algorithm terminates and X_p is presented as the solution. If any variable is changed more than 0.001, X_p is used as the starti.ng point for the next iteration and the algorithm returns to Step I. As with all gradient methods, the algorithm performs very well for initial steps. Convergence toward the solution, when constrained by a linear function, is very rapid. When constrained by a combination of nonlinear constraints, however, convergence near the solution can be slow. The algorithm has been coded to allow limitation on the number of iterations as well as a report of the solution at each step to determine if further effort is necessary. A schematic of the non-linear goal programming algorithm is illustrated in Fig 4-20

Figure 4-20: ***Schematic of the non-linear goal programming algorithm*** ⟶

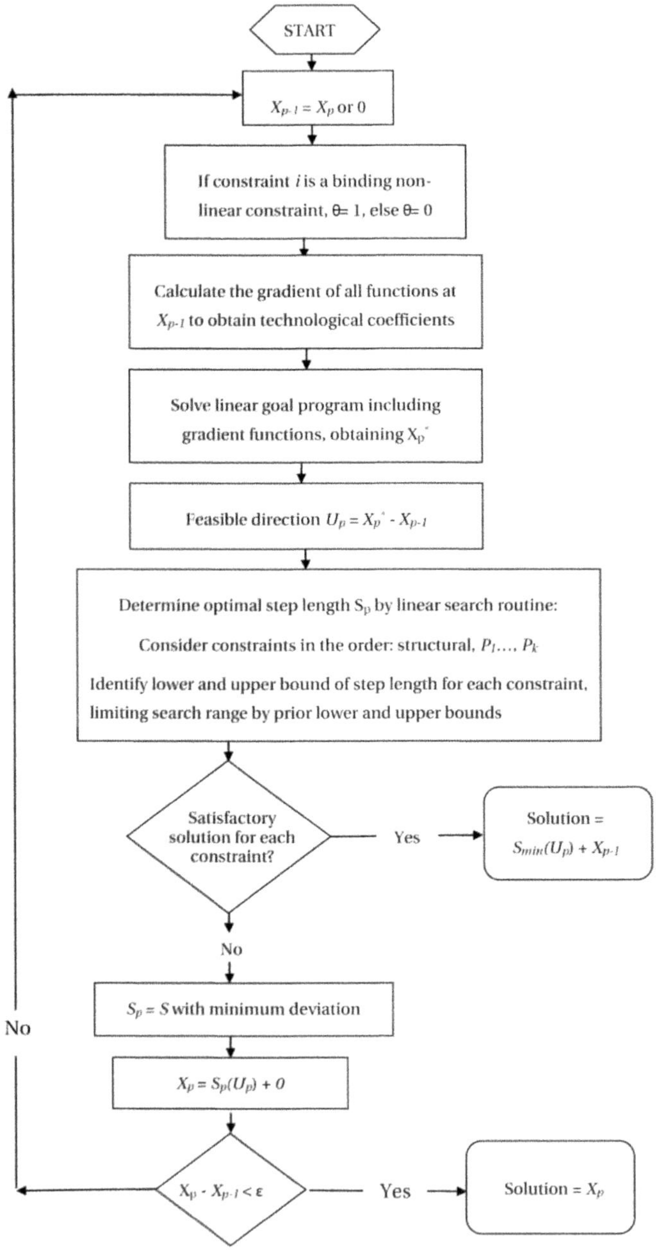
START
Xp-1 = Xp or 0
If constraint i is a binding non-linear constraint, θ= 1, else θ= 0
Calculate the gradient of all functions at Xp-1 to obtain technological coefficients
Solve linear goal program including gradient functions, obtaining Xp*
Feasible direction Up = Xp* - Xp-1
Determine optimal step length Sp by linear search routine:
Consider constraints in the order: structural, P1..., Pk
Identify lower and upper bound of step length for each constraint, limiting search range by prior lower and upper bounds
Satisfactory solution for each constraint?
Yes
Solution = Smin(Up) + Xp-1
No
Sp = S with minimum deviation
Xp = Sp(Up) + 0
Xp - Xp-1 < ε
Yes
Solution = Xp
No

4.14.4 Example

A model is presented to demonstrate the impact of chance constraints in a single objective setting. The model involves the selection of four mix materials so that an optimal design of cattle feed mix can be achieved, and which is subject to protein and fat constraints, as well as having the objective of minimising cost. Data was obtained for the protein and fat content of each material. However, since only the protein constraint was probabilistic in nature, variance data was only obtained for protein. Table 4-42 presents this data.

Table 4-42 Feed mixture data for the work of Lee and Olson (1985)

Variable		Protein Content	Protein Variance	Fat Content	Cost/ton
X_1	Barley	12.0	0.2809	2.3	24.55
X_2	Oats	11.9	0.1936	5.6	26.75
X_3	Sesame Flakes	41.8	20.2500	11.1	39.00
X_4	Groundnut Meal	52.1	0.6241	1.3	40.40

The deterministic equivalent of this model is:

Min cost = $24.55X_1 + 26.750 + 39.00X_3 + 40.50X_4$.

s.t: $X_1 + X_2 + X_3 + X_4 = 1$,

$2.3X_1 + 5.6X_2 + 11.1X_3 + 1.3X_4 \geq 5$,

$12.0X_1 + 11.9X_2 + 48.1X_3 + 52.1X_4 - Z[0.2809X_1^2 + 0.t936X_2^2 + 20.25X_3^2 + 0.6241X_4^2) \geq 21$,

, where Z is the selected degree of confidence desired for satisfying the probabilistic protein constraint. In order to demonstrate the use of goal programming, it is assumed that management would like to balance the cost with the risk of attaining the target amount of protein, and with the design of a mixture containing all four available materials. This would require management to consider target levels for each objective, as well as ranking them by priority. One possible set of objectives is:

Priority 1: Include at least trace elements of all materials.

2. Obtain at least 0.80 confidence that protein $\geq$ 21%.

3. Keep cost $\leq$ 30/unit.

4. Obtain at least 0.90 confidence that protein $\geq$ 21%.

5. Keep cost < 29.90/unit.

6. Obtain at least 0.95 confidence that protein $\geq$ 21%.

7. Keep cost $\leq$ 29.80/unit.

8. Obtain at least 0.98 confidence that protein $\geq$ 21%

9. Minimize cost.

The model for this set of objectives, subject to structural constraints that the four available materials comprise the unit of product and that the fat content be greater than or equal to five units.

Min $Z = P_1(d_1 + d_2^- + d_3^- + d_4^-):P_2d_5 :P_3d_6 :P_4d_7 :P_5d_8 :P_6d_9 :P_7d_{10} :P_8d_{11} :P_9d_{12}$

s.t: $X_1 + X_2 + X_3 + X_4 = 1,$

$2.3X_1 + 5.6X_2 + 11.1X_3 + 1.3X_4 \geq 5,$

$X_1 + d_1 - d_1^- = 0.01,$

$X_2 + d_2 - d_2^- = 0.01,$

$X_3 + d_3 - d_3^- = 0.01,$

$X_4 + d_4 - d_4^- = 0.01,$

$12.0X_1 + 11.9X_2 + 41.8X_3 + 52.1X_4 - 0.841(0.2809X_1^2 + 0.1936X_2^2 + 20.25X_3^2 + 0.6241X_4^2) + d_5^- - d_5^- = 21,$

$24.55X_1 + 26.75X_2, + 39.00X_3 + 40.50X_4 + d_6^- - d_6^- = 30.$

$12.0X_1 + 11.9X_2 + 41.8X_3 + 52.1X_4 - 1.282(0.2809X_1^2 + 0.1936X_2^2 + 20.25X_3^2 + 0.6241X_4^2) + d_7^- - d_7^- = 21,$

$24.55X_1 + 26.75X_2, + 39.00X_3 + 40.50X_4 + d_8^- - d_8^- = 29.9.$

$12.0X_1 + 11.9X_2 + 41.8X_3 + 52.1X_4 - 1.645(0.2809X_1^2 + 0.1936X_2^2 + 20.25X_3^2 + 0.6241X_4^2) + d_9^- - d_9^- = 21,$

$24.55X_1 + 26.75X_2, + 39.00X_3 + 40.50X_4 + d_{10}^- - d_{10}^- = 29.8$

$12.0X_1 + 11.9X_2 + 41.8X_3 + 52.1X_4 - 2.323(0.2809X_1^2 + 0.1936X_2^2 + 20.25X_3^2 + 0.6241X_4^2) + d_{11}^- - d_{11}^- = 21,$

$24.55X_1 + 26.75X_2, + 39.00X_3 + 40.50X_4 + d_{12}^- - d_{12}^- = 0$

a) First pass at the solution of this model would be as follows:

Step 1: *Determine normalized constraints.* At the current solution [0,0,0,0], none of the four chance constraints are equal to the fight hand side value of 21. Therefore, 0 is 0 in all four cases. The chance constraints are normalized with respect to the current solution. In this iteration, the current solution is the origin, resulting in use of the expected a,. Therefore, all chance constraints are:

$12.0X_1 + 11.9X_2 + 41.8X_3 + 52.1X_4 + d_1^- - d_1^- = 21,$

Step 2: *Determine an improving direction.* Solution of the linear goal program with the normalized constraints obtained in step 1 would be:

$X_1'' = [0.65000, 0.05127, 0.28873. 0.01000].$

This solution vector is determined by the two structural constraints, both of which are binding, as well as the linearised chance constraint for protein at priority I (also binding). The minimum limit for x4 is binding. Cost subject to these limits is minimized at 28.99.

Step 3: *Determine optimal direction.* Optimal direction $U_1 = X_1^* - X_0 =$ [0.65000, 0.05127, 0.28873, 0.01000].

Step 4: *Determine optimal step length.* Lower and upper bounds for the step length are determined by sequentially considering constraints in priority order, with structural constraints first. One structural constraint is that the sum of the variables equal t. This is satisfied only by a step length S_1 of 1.0. This step length also yields a binding fat requirement of 5 units.

$X_3 = 1.01[0.65000, 0.05127, 0.28873, 0.01000] + [0, 0, 0, 0].$

Step 5: *Evaluate for convergence.* X_1 is compared with X_0 for all four variables. If the difference for any of the four variables is greater than the parameter t (set to 0.001), a further iteration is required. X_1 differs from X_0 by more than 0.001 for all four variables.

b) Iteration

Step 1: *Determine normalized constraints.* The functional values of the chance constraints evaluated at X_1 are sown in Table 4-43. As none of the chance constraints are binding within the tolerance limit 0.001, θ is 0 for all chance constraints. The gradient for each chance constraint is now substituted for the a_{1i} elements. The general formulation for the gradient of chance constraints with stochastic a_{1i}^+:

Table 4-43 Iteration: Functional values of chance constraints for the work of Lee and Olson (1985)

Constraint	*Z*	$E(\Sigma a_{ij}x_j)$	$\Sigma a_{ij}x_j + Z(Var)^{1/2}$
5	-0.841	21.0003	19.86940
7	-1.282	21.0003	19.27652
9	-1.645	21.0003	19.78850
11	-2.323	21.0003	17.87701

$$\text{New } a_{1i}^+ = a_{1i} + \frac{Z(\Sigma_{k=1}^{n} x_j \text{Variance}_{jh})}{[\Sigma_{j=1}^{n}\Sigma_{m=1}^{n} x_1(\text{Variance}_{jm})x_m]^{1/2}}$$

The coefficients for the protein constraint will now be a function of the new solution X_1 as well as the *Z* value reflecting desired probability of constraint satisfaction. For X_1, $[\Sigma_{j=1}^{n}\Sigma_{m=1}^{n} x_1(\text{Variance}_{jm})x_m]^{1/2}$ = 1.34439, the gradients of each chance constraint are developed for a linear goal programming model as in Table 4-44

Table 4-44 Iteration: - Gradients of chance constraints for the work of Lee and Olson (1985)

Z	Constraint
-0.841	$11.886x_1 + 11.894x_2 + 38.142x_3 + 52.096x_4 + d_5^- + d_5^+ = 21$
-1.282	$11.826x_1 + 11.891x_2 + 36.225x_3 + 52.094x_4 + d_7^- + d_7^+ = 21$
-1.645	$11.777x_1 + 11.888x_2 + 36.646x_3 + 52.093x_4 + d_9^- + d_9^+ = 21$
-2.323	$11.685x_1 + 11.883x_2 + 36.697x_3 + 52.089x_4 + d_{11}^- + d_{11}^+ = 21$

Step 2: *Determine improving direction.* The linear goal programming model including the gradient constraints from Step 1 yields the solution,X_2^* = [0.62747, 0,01000, 0.30914, 0.05339}, The structural limits are binding with the sum of variables equal to 1 and fat content equal to 5.0. Cost is 29.89.

Step 3: *Determine optimal direction.*

$U_2 = X_2^* - X_1$ = [- 0.02253, - 0.04127, 0.02041, 0.04339].

Step 4: *Determine optimal step length.* Step limits are determined for all constraint in priority order. See Table 4-45

Table 4-45 Upper and lower limits for constraints for the work of Lee and Olson (1985)

Constraint		Lower Limit	Upper Limit
	Sum of variables	0.0	1.0
	Fat content	0.0	1.0
P_1	All variables included	0.0	1.0
P_2	Protein probability $\geq$ 0.8	0.496	1.0
P_3	Cost $\leq$ 30	0.496	1.0
P_4	Protein probability $\geq$ 0.9	0.769	1.0
P_5	Cost < 29.9	0.769	1.0
P_6	Protein probability $\geq$ 0.95	1.0 Minimises deviation	

Priority 1 was satisfied by all step lengths between 0 and 1. Priority 2 was satisfied b v all step lengths between 0.496 and l. Therefore, all subsequent searches are limited to this range. Priority 4 further restricted the range. Priority 6 was never satisfied, but the minimum deviation was obtained at a step length of 1.0 and therefore that step length was selected.

Step 5: *Evaluate for convergence.* X_2 is $S_2(U_2) + X_{1^*}$, or 1.0[-0.02253, -0.04127, 0,02041, 0.04339] + [0,65000. 0.05127, 0.28873, 0.010000], X_2, = [0.62747, 0.01000, 0.309!4, 0.05339]. All four variables have changed by more than 0.001. Consequently, another iteration is required.

c) Iteration 3

Step 1: *Determine normalized constraints.* The functional values of the chance constraints evaluated at X_2 are shown in Table 4-46. As none, of the functional values are within 0.001 of the b, of 21, θ = 0 for all four chance constraints, new technological coefficients are developed for the chance constraints by calculating the gradients with respect to X_2

Table 4-46: Iteration 3 – Functional values of chance constraints for the work of Lee and Olson (1985)

Constraint	Z	$E(\Sigma a_{ij}x_j)$	$\Sigma a_{ij}x_j + Z(\text{Var})^{1/2}$
5	-0.847	23.35231	22.14888
7	-1.282	23.35231	21.51782
9	-1.645	23.35231	20.99839
11	-2.323	23.35231	20.02820

Step 2: *Determine improving direction.* The linear goal programming model including the constraints obtained in Step 1 yields X_3^* = [0.62743, 0.01000, 0.30914, 0.05343]. This solution is binding structural constraints. The cost of the solution is 29.89. All goals of the linear model were satisfied until Priority 8, which was to keep cost below 29.80.

Step 3: *Determine optimal direction.* $U_1 = X_3^* - X_2$ [-0.00004, 0, 0, 0.00004]

Step 4: *Determine optimal step length.* The full range of step values from 0 to 1 satisfies the structural constraints as well as goal constraints from Priority 1 through Priority 5. Priority 6 is not satisfied until the step length is 1, determining S_1 = 1.0. X_3 is $S_2(U_2) + X_2$, = [0.62743, 0.01000, 0 30914, 0.05343].

Step 5: *Evaluate for convergence.* See Table 4-47. Since all variables changed less than 0.001, the algorithm terminates with X_3 as the solution. The solution satisfies the structural requirements that the sum of variables equal 1 and that the expected fat content is at least 5 units. The resulting probability of protein content being at least 21 units is 0.95, because that goal constraint was binding for the final solution. The cost of this solution is 29.89. This does not satisfy the 7-th priority objective, but the goal given Priority 6 precludes further satisfaction of cost goals.

Table 4-47: Solution convergence for the work of Lee and Olson (1985)

Variable	X_1	X_2	Difference
x_1	0.62743	0.62747	0.00004
x_2	0.01000	0.01000	0
x_3	0.30914	0.30914	0
x_4	0.05343	0.05339	0.00004

Decision makers could view the solution to their initial set of objectives, and if features of the resulting solution was undesirable, could easily rearrange priority levels, or insert new goals to reflect the added knowledge of the initial run. The algorithm input would require minimal changes to reflect a new order of priorities.

4.14.5 Extraction of the Optimisation Technique for chance constrained non-linear goal programming

Table 4-48 Optimisation under chance constrained goal programming – based on the work of Lee and Olson (1985)

1 Formulate a general linear goal programming problem (for a supply chain environment)

Min $Z = P_1(\sum w_i d_i^{+/-});\ldots; P_k(\sum w_i d_i^{+/-})$;

$$\text{s.t. } \sum a_{ij}x_j + d_i^- - d_i^+ = b_i,\ i = 1,\ldots, m$$

$$b_i, d_i^+, d_i^- \geq 0,\quad i = 1,\ldots,m$$

$$x_j \geq 0,\quad j = 1,\ldots, n$$

, where x_j = j-th decision variable

b_i = target for i

a_{ij} = coefficient of x_j in the i-th goal constraint

d_i^- = underachievement of goal i

d_i^+ = overachievement of goal i

P_k = k-th pre-emptive priority level

w_i = weight assigned to d_i^+ and d_i^- at priority k

2 Introduce chance constraints, e.g. $\Pr(\sum a_{ij}x_j \leq b_i) > \alpha_1$, that yield deterministic equivalents that are non-linear, $E(\sum a_{ij}x_j) + F^{-1}(\beta)[V(\sum a_{ij}x_j - b_i \leq E(b_i)$, where E( ) and V( ) are the expected value and variance operators respectively

3 The gradient algorithm for non-linear goal programming

- Determine normalised constraints
- Determine an improving direction, X_p^*
- Determine optimal direction for the iteration, $U_p = X_p^* - X_{p-1}$
- Determine optimal step length, S_p , $X_p = S_p(U_p) + X_{p-1}$
- Evaluate for convergence (if any variable has changed by more than 0.001 go back to a))

4 Record the optimum solution point, X_p

4.14.6 Conclusion

The gradient nonlinear goal programming algorithm provides an effective means to consider chance constrained models involving multiple goals. The ability to consider this class of decision model expands the support available to organizational decision makers. Consideration of the trade-off between opportunity and risk allows more realistic modelling of decision problems. Consideration of risk in linear models requires surrogate model constraints, such as the imposition of diversity or minimization of average variance. Use of chance constrained programming model allows more direct assessment of the impact of risk levels. Use of goal programming allows consideration of risk level as a goal, while seeking improved expected performance at the same time. Strengths of the algorithm include minimal model preparational effort. The most severe limitation of the algorithm is its specificity to a restricted class of model, as well as the limitations of all gradient based methods when highly nonlinear constraints are binding. The effective optimisation (goal programming) technique utilised therein was extracted and is recorded in Table 4-48.

4.15 Optimisation of a Supply chain Network under Stochastic Fuzzy uncertainty –

- based on the work of Santoso et al. (2005) -

4.15.1 Introduction

It was evident, based on supply chain under uncertainty research conducted by Santoso et al. (2005) that prior stochastic programming research efforts into demand uncertainty were limited by the very small number of probabilistic scenarios that were utilised for modelling purposes. Therefore, in the paper by Santoso et al. (2005), it was decided to extend such previous research by increasing the demand sampling space. This would be achieved by providing an infinite number of demand scenarios through the integration of a sampling strategy, SAA – Sample Average Approximation method, with an accelerated Benders decomposition algorithm to solve supply chain design problems with continuous distributions for the uncertain parameter, and thereby providing for an infinite number of demand scenarios.

For the computational case study, at the end of this paper, it was decided to apply this integrated algorithm in two supply chain networks described by a finite, but an extremely large, number of demand scenarios.

4.15.2 Problem Description

A deterministic mathematical model is formulated by Santoso et al. (2005) to initially describe the supply chain design problem and then to extend it to incorporate the stochastic processing and transportation modelling requirements. The nomenclature requirements for the model are described in Table 4-49

Table 4-49: Nomenclature requirements for supply chain under stochastic uncertainty modelling - Santoso et al. (2005)

Indices	Physical Meaning
i, j, l	nodes
k	products
Parameters	Physical Meaning
G	supply chain network
N	set of nodes
A	set of arcs
S	set of suppliers
P	set of processing facilities
C	set of customers
M	set of manufacturing facilities
F	set of finishing facilities
W	set of warehouses
K	set of products flowing through supply chain
Variables	Physical Meaning
x^k	product flowrate of product, k
c_i	investment cost for building facility i
q_{ij}^k	unit cost for processing product k at facility i and/or transporting product k on arc (i, j)
r_{ij}^k	processing requirement for product k at node j
m_j	the capacity of facility j
Binary Variables	Physical Meaning
y	processing facility establishment decisions, 1 = yes, 0 = no

4.15.3 Model Formulation

The requirement is to minimise total investment and operational costs, therefore:

$$\min \sum_{i\in P} c_i y_i + \sum_{k\in K}\sum_{(i,j)\in A} q_{ijj}^{k}\, x_{i,,j}^{\,k} \qquad (1)$$

subject to: $y \in Y \subseteq (0, 1)$

(local dependencies and restrictions)

$\sum_{i\in N} x_{i,,j}^{\,k} - \sum_{l\in N} x_{j,l}^{\,k} = 0, \quad \forall j \in C, \forall k \in K$

(flow conservation of product k through each processing node)

$\sum_{i\in N} x_{i,,j}^{\,k} \geq d_j^k, \quad \forall j \in C, \forall k \in K$

(total flow of product k to a customer node j should exceed demand d_j^k at that node)

$\sum_{i\in N} x_{i,,j}^{\,k} \leq s_j^k, \quad \forall i \in S, \forall k \in K$

(total flow of product from a supplier node j should be less than the supply s_j^k at that node)

$\sum_{k\in K} r_{jj}^{k} \{\sum_{i\in N} x_{i,,j}^{\,k}\} \leq m_j y_j, \quad \forall j \in P$

(enforces capacity constraints at the nodes)

For the extension of the model above to address stochastic requirements, it is assumed that processing and transportation costs, demands, supplies and capacities are stochastic parameters with known joint probability distribution. However, since it may be impossible to meet demand for certain realisations of uncertain parameters, an additional cost term is included to penalise shortfall. The resulting formulation is as follows:

$$\text{Min } f(y) = \sum_{i \in P} c_i y_i + \text{E}[\text{Q}(y, \xi)] \qquad (2)$$

$$\text{s.t. } y \in Y \subseteq (0, 1)$$

, where Q(y, ξ) is the optimal value of the following problem:

$$\min \quad (\sum_{k \in K} \sum_{(i,j) \in A} q_{ij}^{k} x_{i,j}^{k} + h^{T} z)$$

$$\text{s.t.} \quad \sum_{i \in N} x_{i,j}^{k} - \sum_{l \in N} x_{j,l}^{k} = 0, \qquad \forall j \in C, \forall k \in K$$

$$\sum_{i \in N} x_{i,j}^{k} \geq d_j^{k}, \qquad \forall j \in C, \forall k \in K$$

$$\sum_{i \in N} x_{i,j}^{k} \leq s_i^{k}, \qquad \forall i \in S, \forall k \in K$$

$$\sum_{k \in K} r_{jj}^{k} \{\sum_{i \in N} x_{i,j}^{k}\} \leq m_j y_j, \qquad \forall j \in P$$

, where $h^T z$ corresponds to the penalty, referred to above, incurred for failing to meet demand. This model is now a two stage stochastic program, in which the overall objective is to minimise the current investment costs plus the *expected* future costs of which the first stage involves the determination of the network configuration and then the second stage involves the determination of optimum operating conditions.

According to Santoso et al. (2005), there are two potential difficulties in solving (2). Firstly, this is because an evaluation of the objective function, $f(y)$, for a given configuration, y, requires the determination of the expected value of the linear programming function, i.e. E[Q(y, ξ)], and for continuous distributions this is quite impossible since multiple integrals are involved. Moreover, for discrete distributions, computing the expected value might involve solving a prohibitively large number of linear programs, i.e. one for each scenario. Secondly, even if the expected value function may be computed, the optimisation thereof can presents significant difficulties. Therefore solving (2) involves the minimisation of an implicitly defined non-linear function with respect to binary variables, which can be quite difficult and consequently, a solution algorithm is proposed.

4.15.4 Algorithmic solution method

This solution method involves the integration of a sampling technique, SAA – Sample Average Approximation, with an accelerated Benders decomposition scheme to enable the rapid computation of high quality solutions for large-scale stochastic problems.

a) SAA approximation

i) Generate N realisations (scenarios) of the random vector, ξ

ii) Approximate the expectation vector, $E[Q(y, \xi)]$, by application of the SAA function, $N^{-1}\sum^{N}_{n=1} Q(y, \xi)$

iii) Consequently, such SAA approximation is minimised as follows:

iv) $\min \{\sum_{i \in P} c_i y_i + N^{-1}\sum^{N}_{n=1} Q(y, \xi)\}$ (Master problem)

v) When the sample size, N, is large, it can be shown that the solution of the Master problem is convergent to an optimal solution of probability, one, and, being deterministic, it can be solved by appropriate optimisation techniques.

vi) Such minimum sample size, N, is $N \geq (3\sigma^2_{max})/(\varepsilon - \delta)^2 \log(|Y|/\alpha)$

vii) This relationship guarantees that the SAA solution will be optimal

b) Benders decomposition

The Benders decomposition improves the approximation of $N^{-1}\Sigma^{N}{}_{n=1}$ $Q(y, \xi)$ through an iterative process

4.15.5 Extraction of the programming approach for a supply chain under stochastic demand uncertainty

Table 4-50: Extraction of the programming technique to establish demand scenarios necessary for an optimum solution in a supply chain under stochastic demand uncertainty environment based on the work of Santoso et al. (2004)

1	**Formulate a technique in order to determine the number of demand scenarios necessary for the determination of an optimum solution in a supply chain under stochastic uncertainty environment,**
2	SAA – Sample Average Approximation - Generate N realisations (demand scenarios) of the random vector, ξ, where E[Q(y, ξ)] = Σk∈KΣ(i,j)∈A qijjk xi,,j k - Approximate the expectation vector, E[Q(y, ξ)], by application of the SAA function, N-1ΣN n=1 Q(y, ξ) - Consequently, such SAA approximation is minimised as follows: min {Σi∈P ciyi + N-1ΣN n=1 Q(y, ξ)} (Master problem) When the sample size, N, is large, it can be shown that the solution of the Master problem is convergent to an optimal solution of probability, one, and, being deterministic, it can be solved by appropriate optimisation techniques. - Such minimum sample size, N, is N > (3σ2max) / (ε - δ)2 log(|Y|/α)
3	This technique guarantees that the number of probabilistic scenarios generated will provide an optimum solution

4.15.6 Conclusion

In the paper by Santoso et al. (2005), two different techniques, SAA - Sample Aggregate Approximation, and Benders Decomposition, were discussed that can approximate the number of probabilistic demand scenarios necessary for the determination of an optimum solution in a probabilistically demand uncertain supply chain environment. Both techniques require iterative algorithms for solution and in the case of Benders decomposition, the number of iterations required for solution were discovered to be too large in practise.

The supply chain, under uncertainty, optimisation technique utilised in this procedure was extracted and is recorded in Table 4-50.

Chapter 5

INTEGRATION TO CREATE A PLANNING AND OPTIMISATION METHODOLOGY FOR A SUPPLY CHAIN UNDER UNCERTAINTY

5.1 Introduction

Following the extraction of all the 'Supply Chain, under Uncertainty, Optimisation' techniques utilised in the various selected and analysed research journals, it now remains to logically integrate them so that a generic, 'Supply Chain under prevailing conditions of planning and operational Uncertainty (multi-objectivity, fuzziness, stochastic uncertainty) Optimisation methodology' may be created for any uncertain supply chain environment. By 'logical' is meant being in accordance with the integration procedures utilised by some of the constituent techniques, e.g. the 2SMP technique of Gupta and Maranas (2003), without contravening any potential conditions or limitations.

5.2 Methodology Analysis

A comparison between all the various extracted techniques can best be achieved by viewing Tables 5-1, which presents all techniques in a numbered and structurally consistent format. The development of such structurally consistent format was based on an analysis of all extracted techniques and then the identification of all possible and important sequential steps. The identification of all possible and important sequential steps was vital to ensure that a generic 'Supply Chain, under prevailing conditions of uncertainty, Optimisation' methodology could be developed.

Tables 5-1: Structurally consistent supply chain, under uncertainty, techniques

	Bit et al. **Multi-obj. Fuzzy Opt**	**Gupta & Maranas** **Multi-obj. Stochastic Opt.**	**Tsiakis &Papageorgiou** **Multi-obj. Fuzzy Opt**
1	Describe the supply chain environment	Describe the supply chain network	Describe the supply chain network environment
2			Illustrated diagram of SCN
3		Identification of binary production/network variables	Identification of binary production/network variables
4	Requirements Definition	Requirements Definition	Requirements Definition
5	Multi-objective Fuzzy MILP programming plus Asp. Level subroutine for max-min scenarios	Bi-level stochastic programming to accommodate two phases, manufacturing and logistics	Multi-objective Fuzzy MILP programming with different scenario configurations
6	Nomenclature	Nomenclature	Nomenclature
7	Formulate a MILP (Mixed Integer Linear Program0 - Define objectives - Specify constraints and limitations	Formulate a two-stage stochastic MINLP (2SMP) program - First-stage, 'variables are determined prior to resolution of the - Second-stage, variables - Define all constraints	Formulate a MILP program - defining the objectives - define all constraints
8	Execute the Aspiration level program to determine, max-min scenarios	Coupling between the 1st and 2nd stages is achieved by constraint variable resolution	
9	Solve the model and record the optimal solution	This recourse optimisation problem determines optimal 2nd stage results given the optimal 1st stage results	Solution of the supply chain network MILP model based on a number of possible configuration scenario's

	Chen et al. Multi-obj. Stoch. Fuzzy Opt	**Grossmann and You Multi-obj. Fuzzy Stoch. Opt.**	**Guillen et al. Multi-obj. Fuzzy Stoch. Opt.**
1	Description of the SCN	Description of the SCN	Description of the SCN
2	Illustrated diagram of supply chain network	Illustrated diagram of supply chain network	Illustrated diagram of supply chain network
3	Identification binary prod./network variables	Identification binary prod./network variables	
4	Requirements Definition	Requirements Definition	Requirements Definition
5	Discrete prob. Scenarios followed by fuzzy optimisation	MINLP modelling with probabilistic constraints	Discrete stochastic scenario's followed by MINLP program
6	Nomenclature	Nomenclature	Nomenclature
7	Formulate a Mixed Integer Non-Linear Programming model (MINLP) - Define problem and deterministic objectives - Define all probabilistic and deterministic constraints - Convert prob. obj. and constr. into deterministic equivalents	Formulate a MINLP or MILP program. - Define multiple objectives - Specify all constraints for each probabilistic scenario sub-program - Convert all probabilistic objectives and constraints into deterministic equivalents	Formulate a MINLP program - Define multiple obj. - Defining all deterministic and probabilistic constraints - Convert all probabilistic objectives and constraints into deterministic ones
8	Compromise soln. obtained:	Solve the MINLP program per scenario	Solve the MINLP program
9	Compute average result for each probabilistic objective over scenario's using expectation operators.	Generation of optimum results by summating all various weighted probabilistic scenario's	Compute the final result by the summation of prob. Weighted decision variables

	C-Valdés and Ozdemir (2011) **Bi-obj. Stoch. Fuzzy Opt**	**Razmi et al. (2013)** **Bi-obj. Stoch. Fuzzy Opt.**	**Baky (2010)** **Multi-obj. Fuzzy Opt.**
1	Description SCN	Description of SCN	Description of SCN
2	Identification binary prod./network variables	Identification binary prod./network variables	
3	Requirements Definition	Requirements Definition	Requirements Definition
4	Stochastic demand uncertainty with MILP modelling with fuzzy production and distribution uncertainty	Demand uncertainty is represented by a number of different probabilistic scenarios and fuzzy MILP solution approach	Multi-objective fuzzy uncertainty with objective functions transformed into membership functions, which then allow the formation of a MILP model
5	Nomenclature	Nomenclature	Nomenclature
6	Formulate a Mixed Integer Linear Programming model (MILP) - Define prob. and deter. objectives & constr. - Convert all probabilistic obj. and constraints into deterministic equiv.	Formulate a MINLP program. - Define multiple objectives - Specify all constraints for each probabilistic scenario - Convert all probabilistic objectives and constraints into deter. equiv.	Elicit the membership functions per objective function - Set the goals and limits per obj. Fn. - Elicit the membership functions.
7	Transform one objective function into a Pareto ε-constraint	Transform one objective into a Pareto ε-constraint	Transform membership functions into MILP with associated objective functions and constraints
8	Solve the MILP and record the Pareto optimal result	Generation of optimum solution by summating all probabilistic scenario's	Solve the MILP per objective and record the results

	Guillen and Grossmann Multi-obj. Stoch. Fuzzy Opt	**Liu and Papageorgiou Multi-obj. Stoch. Fuzzy Opt.**
1	Description of the supply chain network	Description of the supply chain
2		
3	Identification binary prod./network variables	Identification binary prod./network variables
4	Requirements Definition	Requirements Definition
5	MINLP modelling with multi-objective /fuzzy/stochastic constraints	MILP modelling with multi-objective/fuzzy/stochastic constraints
6	Nomenclature	Nomenclature
7	Formulate a MINLP model - Define probabilistic and deterministic objectives - Define all probabilistic and deterministic constraints - Convert all probabilistic obj. and constraints into deterministic ones - Integrate the environmental impact model	Formulate a MILP model - Define multiple objectives - Specify all constraints for each probabilistic scenario sub-program - Convert all probabilistic objectives and constraints into deterministic equivalents
8	Transform one objective into a Pareto ε-constraint	Transform one objective into a Pareto ε-constraint and solve the MILP program per scenario
9	Solve the MINLP	Solve the MILP per scenario and obtain optimum solution by summating all probabilistic scenario's

Now that a numbered and structurally consistent format for all analysed supply chain, under uncertainty, optimisation techniques has been created, it is now becomes possible, in the following section, to identify a common core format and also to identify any necessary exceptions and how those are handled, e.g. ∈-constraint, multi-objective Pareto optimality and transformation of probabilistic relationships into deterministic equivalents.

5.3 Extracted Technique Integration

.An analysis of the extracted Optimisation under Uncertainty techniques from the research papers listed in Tables 5-1 reveals that there is a common theme to the basic structure of those techniques that also caters for any variations, e.g. nature of demand uncertainty and solution approach, nature of constraints etc. In addition to this, certain indigenous technique steps/techniques were ignored because they weren't the latest accepted procedures for the process involved, e.g. the Bit et al. (1992) Aspiration Level multi-objective routine that has long been surpassed by the Pareto ε-constraint routine (Guillen and Grossmann (2009) and Liu and Papageorgiou (2013) This common theme structure, with variation, for all extracted methodologies is presented in Table 5-2

Table 5-2 Common theme structure and breakdown of extracted optimisation of supply chain under uncertainty methodologies

Section	**Common constituent sequential components**
Introduction	Description of supply chain network' Description of the problem Illustrated diagram of the supply chain network Identification of all desired supply chain linkages, i.e. production plant (Pl) – distribution centre (Dc) and Dc – Customer Zone (Cz), using binary variables, e.g. x_{ij} and y_{jk} Nomenclature Section
Nature of Uncertainty and Solution Approach	Fuzzy demand (sales $\leq$ x tons per period) followed by MILP/MINLP programming' solution ***Or*** Discrete probabilistic demand scenario's, followed by solution of MILP/MINLP program per scenario and then weighting of optimum results according to respective probabilities ***Or*** Continuous probabilistic demand. Transformation of stochastic relationships into deterministic ones and then solution of MILP/MINLP program ***Or***

	A two stage stochastic modelling program (2SMP) for an inventory level-demand based management system. The two stages (production and logistics) could be any appropriate Uncertainty combination
Supply Chain Modelling, Programming and Execution (this typically represents the bulk of the work)	Dependent on the approach decided above but basically involves the creation of a Linear Program under Uncertainty Model - Formulation of operational objective relationship(s) - Formulation of constraint equation relationships for both the individual supply chain activities, i.e. production and distribution, as well as for the binary network constraints - If probabilistic objectives or constraints exist, then the transformation of these into deterministic equivalents e.g. using techniques of Lee and Olson (1985) ***And /Or Not*** - In the event of multiple objectives, the principal objective is selected as the MILP/MINLP objective whilst the others are allocated as Pareto ε (epsilon) constraints, with the value of ε being either the upper (max) or lower (min) limit of the ε-constraints - Final formulation of (deterministic) MILP/MINLP
Solution Processing	Solution of Linear Program(s) under Uncertainty (typically computer aided, e.g. Microsoft Excel Solver)
Presentation of optimum results	

It is clear from Table 5-2 that the key decision point for the Optimisation of a Supply Chain under Uncertainty is the nature and treatment of Demand Uncertainty. The reported options are:

1) Fuzzy demand followed by MILP/MINLP programming' solution.

2) Discrete probabilistic demand scenario's, followed by solution of MILP/MINLP program per scenario and then weighting of optimum results according to respective probabilities.

3) Continuous probabilistic demand - Transformation of stochastic relationships into deterministic ones and then solution of MILP/MINLP program

4) A two stage stochastic modelling program (2SMP) for an inventory level-demand based management system. The two stages (typically production and logistics) could be any appropriate Uncertainty combination

It is therefore now possible to construct a 'Planning and Optimisation Methodology' for Supply Chains that operate under conditions of Uncertainty, which, based on the contents of Table 5-2, can accommodate a variety of Demand Uncertainty options as well as accommodating some operational variations such as the nature of constraints, be they deterministic or probabilistic.

Chapter 6

SPECIFICATION OF A PLANNING AND OPTIMISATION METHODOLOGY FOR A SUPPLY CHAIN NETWORK UNDER UNCERTAINTY

6.1 Flow Diagram of Derived Methodology

A process flow diagram representation of the 'Planning and Optimisation Methodology for a Supply Chain operating under Uncertainty' can now be presented (Fig 6-1), initially, to be followed by a more detailed tabular description, Table 6-2, of the various stages involved

Figure 6-1: Flow Diagram Description of Methodology

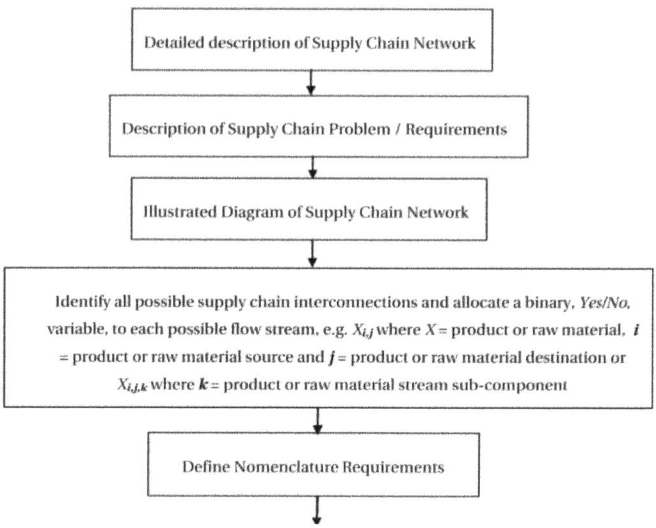

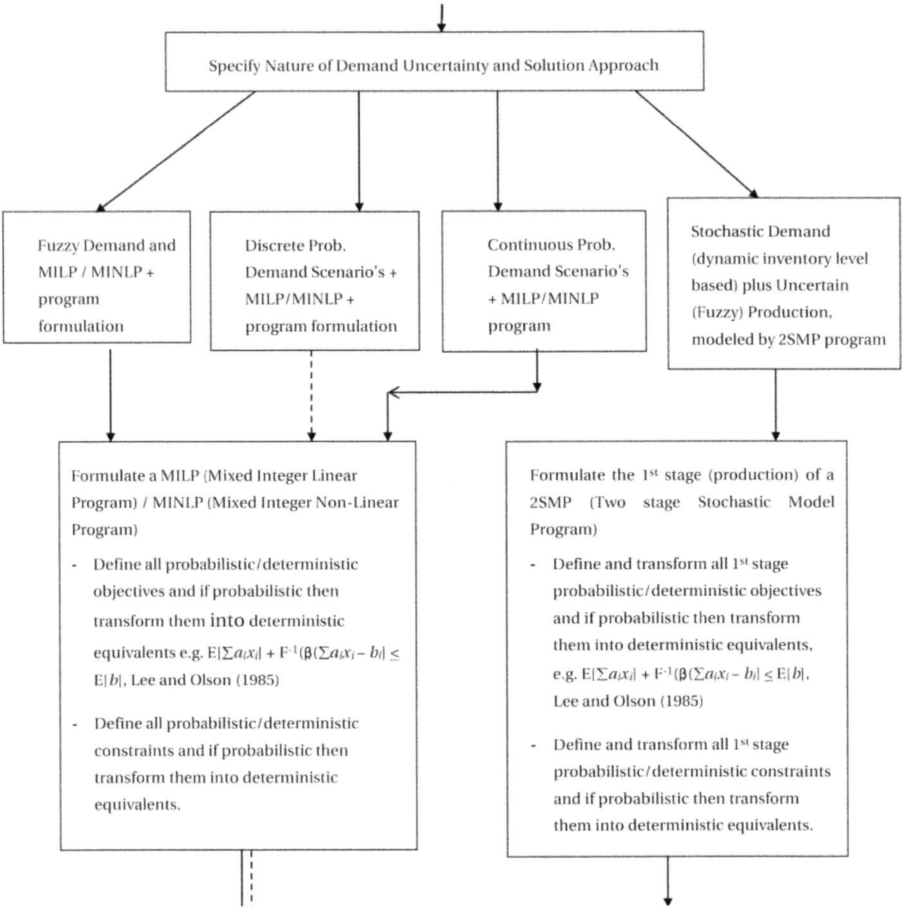
Specify Nature of Demand Uncertainty and Solution Approach
Fuzzy Demand and MILP / MINLP + program formulation
Discrete Prob. Demand Scenario's + MILP/MINLP + program formulation
Continuous Prob. Demand Scenario's + MILP/MINLP program
Stochastic Demand (dynamic inventory level based) plus Uncertain (Fuzzy) Production, modeled by 2SMP program
Formulate a MILP (Mixed Integer Linear Program) / MINLP (Mixed Integer Non-Linear Program)
- Define all probabilistic/deterministic objectives and if probabilistic then transform them into deterministic equivalents e.g. E[Σaᵢxᵢ] + F⁻¹(β(Σaᵢxᵢ – bᵢ] ≤ E[b], Lee and Olson (1985)
- Define all probabilistic/deterministic constraints and if probabilistic then transform them into deterministic equivalents.
Formulate the 1st stage (production) of a 2SMP (Two stage Stochastic Model Program)
- Define and transform all 1st stage probabilistic/deterministic objectives and if probabilistic then transform them into deterministic equivalents, e.g. E[Σaᵢxᵢ] + F⁻¹(β(Σaᵢxᵢ – bᵢ] ≤ E[b], Lee and Olson (1985)
- Define and transform all 1st stage probabilistic/deterministic constraints and if probabilistic then transform them into deterministic equivalents.

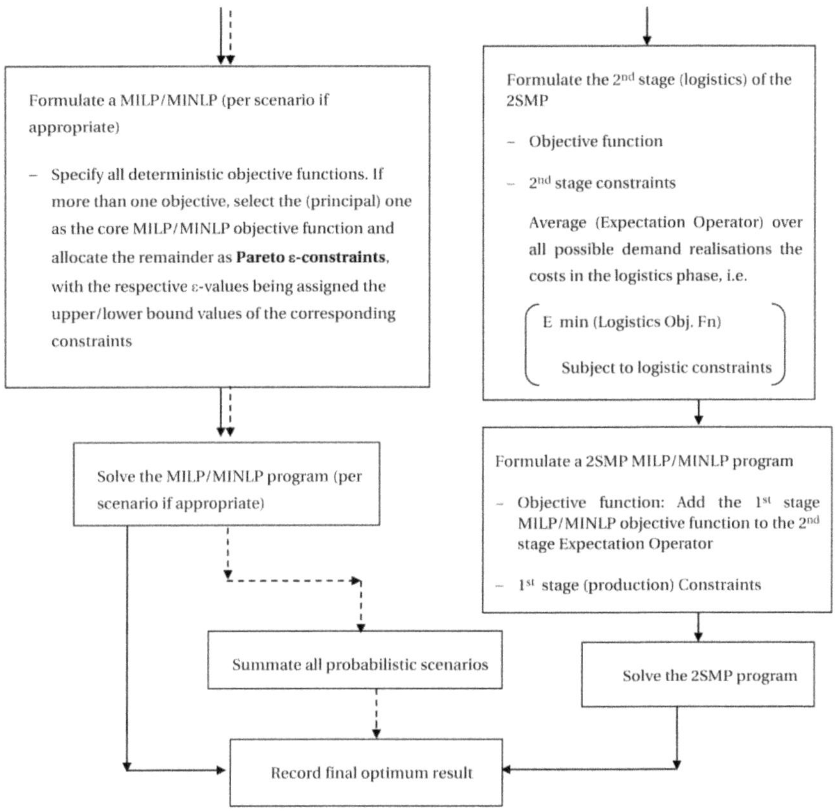

Figure 6-2: Process flow diagram of the 'Supply Chain under prevailing conditions of Uncertainty' Optimisation methodology

6.2 Tabular Description of Methodology Stages

Table 6-1 is purely a stage-wise extraction from the final 'Optimisation of a Supply Chain under prevailing conditions of Uncertainty' methodology derived, and presented in Fig 6-1, but also includes enhanced stage descriptions to ensure greater clarity for application purposes.

Table 6-1 Tabular specification of final 'Planning and Optimisation methodology for a Supply Chain under Uncertainty

Stage	Stage Description
1	Detailed description of Supply Chain Network This must include a description of all supply chain components including all operational performance parameters, e.g. capacity, rate etc.
2	Description of Supply Chain Problem / Requirement These are the operational / infrastructural requirements of the supply chain network of which there could be more than one, e.g. minimum operating cost and/or maximum operating profit etc.
3	Illustrated diagram of Supply Chain Network A diagram of the entire supply chain network including an identification of all key components and all key process flow variables
4	Identify all possible supply chain interconnections and allocate a binary, *Yes/No*, variable, to each possible flow stream, e.g. $X_{i,j}$ where X = product or raw material, i = product or raw material source and j = product or raw material destination or $X_{i,j,k}$ where k = product or raw material stream sub-component
5	Nomenclature Section Definitions of all indices, parameters, variables and constants used
6	Specify nature of Demand Uncertainty and corresponding solution approach This is probably the key decision point of the entire methodology, which determines the subsequent nature of the solution approach. Options are below

7	- Fuzzy Demand and subsequent MILP (Mixed Integer Linear Program) or MINLP (Mixed Integer Non-Linear Program) programming – most supply chain network operations would fall under this category - Discrete Probability Demand Scenarios and subsequent MILP or MINLP programming – Overall product demand could be broken down into a number of scenario's, each with a specific probability of occurrence - Continuous Probability Demand and subsequent MILP or MINLP programming – Product Demand can be represented by probability distribution function, i.e. $F(x)$ - Stochastic Demand and Uncertain Production (typically fuzzy) modelled by a 2SMP (Two-Stage Model Program) program – ideal for supply chain modelling when Demand is primarily determined by inventory level dynamics, e.g. mass market commodities	
	Fuzzy, Discrete and Continuous Probability Product Demand (a)	**2SMP – Two Stage Stochastic Model Programming (b)**
8	Define a MILP or MINLP model depending on the linearity, or not, of the objective functions and constraints - Define all probabilistic and/or deterministic objectives, and if the former, transform them into deterministic equivalents, e.g. $E[\sum a_i x_i] + F^{-1}(\beta)(\sum a_i x_i - b_i] \leq E[b]$, Lee and Olson (1985) - Define all probabilistic and/or deterministic constraints, and if the former, transform them into deterministic equivalents.	Define and formulate a 1st stage 2SMP Stochastic Model - Define and formulate all 1st stage (production) probabilistic and/or deterministic objectives, and if the former, transform them into deterministic equivalents. - Define and formulate all 1st stage probabilistic and/or deterministic constraints, and if the former, transform them into deterministic equivalents.
9	Formulate a MILP or MINLP program (per prob. scenario if appropriate) - Deterministic objective function(s). If more than one objective function, then one (core) objective is selected as the single MILP/MINLP objective whilst the remaining objectives are assigned as Pareto ε-constraints, with the respective	Define and formulate a 2nd stage (logistics) MILP/MINLP - objective function - 2nd stage constraints\ Average (using Expectation Operator) over all possible demand realisations, the

	ε-values corresponding to the upper or lower bound values of the constraints concerned - Deterministic constraints	costs in the logistics phase, i.e. {E min(Logistics Objective) Subj. Logistics onstraints}
10		Formulate a 2SMP program - objective function: add the 1st stage objective function to the 2nd stage Expectation Operator (E) - 1st stage (production) constraints
11	Solve the MILP/MINLP (per scenario if appropriate)	Solve the 2SMP program
12	Summate all probabilistic scenario's (decision variables) to get final optimum result	
13	Record Final Optimal Solution	

6.3 Conclusion

The derivation of a Planning and an Optimisation Methodology for a Supply Chain under Uncertainty has now been achieved and presented in two different formats, i.e. process flowchart and descriptive tabular formats.

As has been mentioned previously, the integrity of each constituent individual and/or combinational optimisation under uncertainty technique has been maintained through the use of an integration technique that recognises and caters for all constituent procedures. It now remains to apply the derived methodology to a live supply chain environment, i.e. the fertiliser division of a large chemical group in South Africa, the results from which will testify to the validity of the methodology.

Chapter 7

Application Example: The Planning and Optimisation of a Supply Chain under Uncertainty

7.1 Introduction

It now becomes necessary to test the derived methodology against an appropriate sample application. By '*appropriate*' is meant a sample application that effectively demonstrates the scope and variability of the derived methodology. This would be fundamental to the study as it would not only demonstrate the application of all the steps of the derived, integrated methodology, highlight the type of difficulties encountered and the suggested methods for dealing with them but also provide an ideal test bed for the derived methodology.

One such comprehensive operating entity has been discovered and that is the production, distribution and sales activities of a large fertiliser and explosives chemical group in South Africa, Such group, which has a comprehensive supply chain network infrastructure both within the country and beyond. is an ideal testing environment for this research because apart from its supply chain network, it has a large integrated fertiliser production facility that produces quite a wide variety of similar product.

Most importantly, the various operations in the abovementioned example incorporate most of the elements of uncertainty, two out of three, in the planning and optimisation of a supply chain under uncertainty, i.e. Multiple Objectivity (minima, maxima or a combination of minima and maxima) and Fuzziness ($<$, $\lesssim$, $>$, $\gtrsim$). Therefore it will certainly be a useful test platform.

7.2 Example Application: A NPK Fertiliser Group - Optimisation of a Supply Chain Network under prevailing conditions of operational Uncertainty

7.2.1 Introduction

A South African Fertiliser Group produces dry, liquid and speciality fertilisers, and has production plants throughout South Africa. The fertiliser division produces and sells plant nutrition products directly to farmers as well as to cooperatives and wholesalers within South Africa. This SA Fertiliser company manages exports into the rest of Africa from Johannesburg, as well as its own facilities in certain African countries. Such African operations have specialised divisions focusing on the needs of entrant farmers – an important developing market for this SA Fertiliser company.

However, the focus here is on the supply chain of one particular range of speciality fertilisers produced, distributed and sold by the SA Fertiliser Group in South Africa, and that is the Nitrogen, Phosphorous, Potassium (NPK) (Appendix 8.1) range of granular fertilisers. The main reason for this is that there are about 30 to 40 different, FSSA, FAO-UN, (2005), blends of NPK fertiliser available, in the South African agricultural sector with each blend geared towards the nutrition of a specific crop type in terms of its NPK ratio, e.g. N:P:K = (2:3:2) with each number representing the relative content of that chemical element in the fertiliser, where N (nitrogen) promotes leaf growth and forms proteins and chlorophyll, P (phosphorus) contributes to root, flower and fruit development and K (potassium) contributes to stem and root growth and the synthesis of proteins. The SA Fertiliser Group only focuses on the production and distribution of NPK fertiliser for the fifteen (15) main crop-types in SA, Table 7-5

Apart from having to produce a whole variety of different NPK-based granular fertilisers that vary according to crop type, which gives rise to product-market uncertainty, there are other operational uncertainties that exist in this commercial operation. These vary from fuzzy fertiliser demand uncertainty, on account of uncertain market demand of NPK fertiliser, to stochastic uncertainty, which arises from probabilistic weather conditions (however, stochastic weather uncertainty has been ameliorated to a large extent by the introduction and existence of sophisticated irrigation systems and therefore will be ignored in this case study), and then to multi-objective uncertainty which arises from the need to maximise a whole range of NPK fertilisers in accordance with market demand, but in conjunction with minimising the discharge of hazardous chemicals from then production process.

However, it is the primary ambition of SA Fertiliser Ltd to maximise the production and distribution of NPK granular fertilisers, in accordance with market demand in an uncertain product and market environment, whilst also minimising the discharge of hazardous effluent. Therefore the optimisation methodology to be followed in this endeavour will be based on the general optimisation methodology for a supply chain under prevailing conditions of uncertainty, as given in Table 7-3. The required plant operating data and technical specifications will be obtained from the corresponding operating and technical manuals on the production facilities of the SA Fertiliser Group.

7.2.2 Detailed description of Supply Chain Network

It is evident from Fig 7-3, a comprehensive flowchart of the entire fertiliser production and distribution operations, that the final fertiliser product production processes, i.e. the various Granulators, are initiated by a number of parallel production processes (i.e. nitric acid, ammonium nitrate solution, nitrophosphate and superphosphate production plants) that constitute the production of the various raw materials required in this NPK fertiliser production. Such parallel raw material production processes, including final fertiliser production, are illustrated in Fig 7-1. All plant operational details were obtained from the various plant technical manuals.

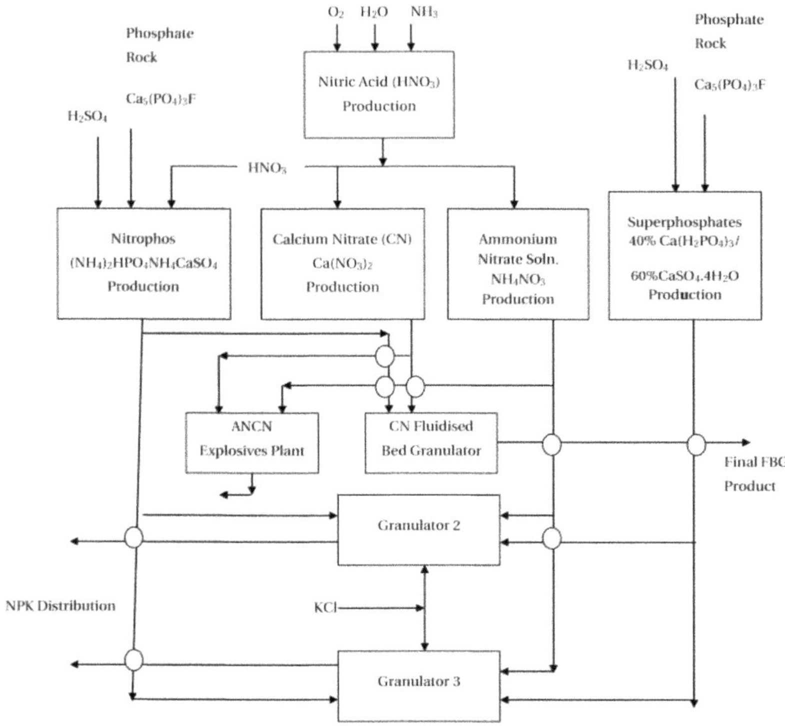

Figure 7-1 ***Simplified flowchart of fertiliser production operations***

The following sections will cover a brief description of the various production processes involved

7.2.2.1 Nitric Acid (NA) Plant

Nitric Acid (HNO_3) is used, in fertiliser production, as an intermediate product for the production of ammonium nitrate solution as well as for the production of ammonium nitrate and calcium based fertilisers.

Air is extracted from the atmosphere at ambient temperature and pressure and is then compressed to a pressure of 4 bar, and at a temperature of 260^{0}C. The compressed air stream is split into two streams, one for the ammonia/air reactor and the other, cooled, for the bleaching tower.

Ammonia is supplied as a gas and fed, along with the air, into the ammonia/air reactor. The catalyst used in the reactor is platinum-palladium gauze. The reactor typically operates at a temperature of 890^0C and at a pressure of 4 bar and the reaction efficiency lies typically between 93% and 98%. The reactions taking place in the reactor are twofold, producing nitrous oxide (NO) and nitrogen dioxide (NO_2) as follows:

$$4NH_3 + 50_2 \rightarrow 4NO + 6H_2O;\ 866.8\ kJ/mole$$

The nitric oxide that is produced at the platinum gauze is oxidised further by means of secondary air reacting homogeneously in the gas phase in accordance with the overall equation:

$$2NO + O_2 \rightarrow N_2O_4$$

Formation of di-nitrogen tetroxide proceeds via an intermediate nitrogen dioxide, NO_2. The nitric oxide oxidation reaction has a negative temperature coefficient hence low temperatures favour high yields. Elevated pressures also raise the yields of the tetroxide, making possible the production of higher strengths of acid.

The hot gas from the reactor is passed through a Lamont boiler where the heat is exchanged with distilled water to produce superheated steam, after which it flows through two energy recovery stages, i.e. the tail gas heater and the economiser.

The gaseous product stream, which, at that stage, is mainly nitrogen dioxide (NO_2) is then fed into the base of the absorption column whilst water and weak nitric are fed into the top of the absorption column. Di-nitrogen tetroxide is absorbed in water with the formation of nitric acid in accordance with the simplified equation:

$$3N_2O_4 + 2H_2O \rightarrow 4HNO_3 + 2NO$$

The absorption proceeds at greater rates, and with higher efficiencies, at elevated pressures and low temperatures. NO_2 is absorbed into the water and eventually 59% nitric acid is formed. Clean gas with only traces of NO is released into the atmosphere as tail gas

7.2.2.2 Ammonium Nitrate (AN) Solution Plant

According to the operating process manual of SA Fertiliser Ltd, ammonia and 59% Nitric Acid are reacted together in a forced circulation loop to produce 89% ammonium nitrate by the following reaction:

$$NH_3 + 59\%HNO_3 \rightarrow 89\%NH_4NO_3 + H_2O$$

For industrial production, this is done using anhydrous ammonia gas and concentrated nitric acid. This reaction is violent and very exothermic. After the solution is formed, typically at about 81% concentration, the excess water is evaporated to an ammonium nitrate (AN) content of between 88% to 95% concentration (AN melt), depending, primarily, on the strength of nitric acid..

The reactor operates at a temperature of 150°C, with a reactor efficiency of between 95% and 97.5%, and the resultant stream is then flashed in the flash chamber. The liquid stream is then circulated back to the reactor while the gas stream is transferred to the scrubbers before being released into the atmosphere. Ammonium Nitrate from the flash chamber is neutralised by ammonia in the neutraliser tank. Ammonium Nitrate, at a concentration of 89%, is stored in two storage tanks. Despatches to external depots are exercised from the storage tanks

7.2.2.3 Liquid Calcium Nitrate (CN) Plant

Liquid CN is produced for use in the solid CN plant as well as for use in the ANCN (Ammonium Nitrate Calcium Nitrate) plant, which delivers a mixture of ANCN ($Ca(NO_3)_2$ and NH_4NO_3 (ammonium nitrate)) for the Blasting Explosives operation.

Liquid CN is produced from a reaction of 59% nitric acid (HNO_3) with lime ($CaCO_3$ - calcium carbonate) Both reactants are both fed into an initial reactor and then overflowed into a secondary reactor for the purpose of increasing overall reactor residence time and thus enhancing the conversion of reactants as follows:

$$CaCO_3 + 2HNO_3 \rightarrow Ca(NO_3)_2 + CO_2 + H_2O$$

Product from the reactor is fed to the primary filters where the sludge is removed and the calcium nitrate solution is then stored in an intermediate tank. From this intermediate tank, calcium nitrate is pumped to a CN storage tank.

If the liquid CN is to be used in the ANCN plant, the pH is first adjusted by means of ammonia and then the liquid CN is pumped to the CN storage tank. If liquid CN is to be used in the solid CN plant, the pH is first adjusted using lime, after which it is fed through secondary filters to produce a clear CN product, which is used as a raw material for the solid CN plant.

7.2.2.4 Nitrophosphate (NP) Plant

The Nitrophosphate production process (Olanipekun (2003)) involves acidifying phosphate rock with nitric acid to produce a mixture of phosphoric acid and calcium nitrate and is shown below

Note: *It is reported by SA Fertiliser Ltd that the phosphate apatite rock that is used in the granular fertiliser production operations is sourced from a northern province in SA and consists of the following combinational mixture of apatite rocks: 60% fluor apatite -* $Ca_{10}(PO_4)_6F_2$*, 40% mixture of hydroxyapatite -* $Ca_5(PO_4)_3OH$ *and carbonic apatite -* $Ca_5(PO_4)_32CO_3$*. For the sake of brevity and convenience in this report, feedstock phosphate apatite rock will, henceforth, be referred to as* $Ca_{10}(PO_4)_6F_2$*, on account of its 60% prevalence of occurrence and also because this is not a critical factor in this research. Additionally, since none of the final granulated NPK fertiliser products contain any -F-, -OH, -*CO_3 *related compounds in their composition; the precise apatite mixture in phosphate rock is irrelevant in any operational mass balance calculations since all such compounds eventually end up as some form of effluent discharge.*

The Nitrophosphate reaction sequence is depicted in Fig 7-2

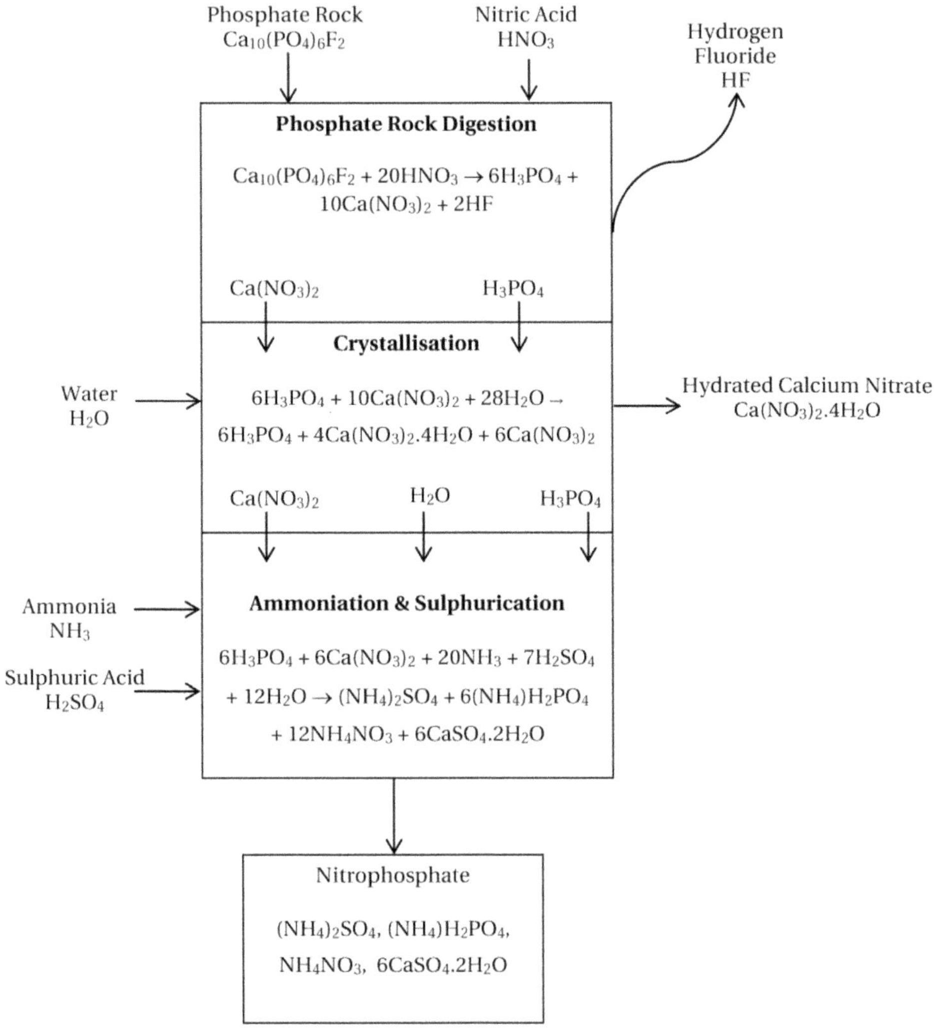

Figure 7-2: *Nitrophosphate Reaction Sequence*

As depicted in Fig 7-2, the reaction sequence in the Nitrophosphate plant is:

1) $Ca_{10}(PO_4)_6F_2 + 20HNO_3 \rightarrow 6H_3PO_4 + 10Ca(NO_3)_2 + 2HF$

The mixture is cooled to below 0 degrees Celsius, where the calcium nitrate crystallises and can be separated from the phosphoric acid. This crystallised calcium nitrate constitutes 70% of the total incoming calcium

2) $2H_3PO_4 + 4Ca(NO_3)_2 + 4H_2O \rightarrow 2H_3PO_4 + Ca(NO_3)_2.4H_2O + 3Ca(NO_3)_2$

The resulting calcium nitrate produces nitrogen fertilizer. The filtrate is composed mainly of phosphoric acid with some nitric acid and traces of calcium nitrate, and this is neutralised with ammonia and reacted with 98% sulphuric acid to produce a nitrophosphate fertiliser mixture consisting of ammonium sulphate ($(NH_4)_2SO_4$) ammonium phosphate ($(NH_4)H_2PO_4$), ammonium nitrate (NH_4NO_3) and hydrated calcium sulphate ($6CaSO_4.2H_2O$), commonly referred to as Nitrophos in the fertiliser industry

3) $6H_3PO_4 + 6Ca(NO_3)_2 + 20NH_3 + 7H_2SO_4 + 12H_2O \rightarrow (NH_4)_2SO_4 + 6(NH_4)H_2PO_4 + 12NH_4NO_3 + 6CaSO_4.2H_2O$

7.2.2.5 *Ammonium Nitrate Calcium Nitrate (ANCN) Plant*

ANCN (Ammonium Nitrate Calcium Nitrate) is produced by mixing 89% Ammonium Nitrate (NH_3NO_4) with 59% Nitric Acid (HNO_3) and then by evaporating off excess water. ANCN is used as an intermediate product in the manufacture of explosives and the mixture is created as follows:

$NH_4NO_4 + Ca(NO_3)_2 \rightarrow NH_4NO_4.Ca(NO_3)_2$

ANCN is produced in batches. Calcium nitrate from the CN plant is transferred to two tanks on the ANCN plant where it is mixed with ammonium nitrate, transferred from the AN plant. The mixture is treated with steam to evaporate water from the mixture until the product has attained the required moisture content.

The product is then transferred to an ANCN storage tank from where it is transferred to explosives manufacturing facilities

7.2.2.6 *CN Fluidised Bed Granulator*

Liquid calcium nitrate (CN) is transferred to one of the holding tanks in the Raw Material Storage section, where ammonium nitrate is added and mixed. The liquid is then concentrated up by use of the steam coils in the concentrator .tanks. A fan is then used to extract the water vapour in this section. The concentrated liquid CN is injected into the fluidised bed granulator via atomising spray nozzles using instrument air. The fluidised bed granulator is supplied with dry air, and the coolers with cold air, in the granulator, and the bed of CN granules is coated with spray. The granules that are formed move through the bed into the fluidised bed cooler before being transferred to the vibrating screens. Hot air from the granulator is scrubbed with a weak CN solution to remove CN dust from the air stream. The weak CN solution is returned to the CN storage tanks and clean air is emitted by the scrubber.

7.2.2.7 *Single Super Phosphate Plant*

Apatite phosphate rock [$Ca_{10}(PO_4)_6$], similarly sourced to that of the Nitrophosphate plant phosphate rock feedstock, is fed though an air swept roller mill where it is ground to $\pm$ 90% less than 53 micrometres. This ground rock is stored in a ground rock hopper from whence it is fed to a mixer on the plant. In the mixer, the ground phosphate rock is reacted with a mixture of 98% sulphuric acid and water according to the following reaction:

$$Ca_{10}(PO_4)_6 + 7H_2SO_4 + 3H_2O \rightarrow 3CaH_4(PO_4)_2H_2O + 2HF(g) + 7CaSO_4$$

The reaction also releases fluorinated gas (HF), which is removed via a scrubber system as follows:

$$4HF(g) + SiO_2 \rightarrow SiF_2(g) + 2H_2O$$

The product from the mixer is then fed through a den, which is a slow moving, fully enclosed conveyor, within which the reaction continues and the emerging product, Single Super Phosphate, $40\%Ca(H_2PO_4)_3$ plus $60\%CaSO_4.4H_2O$, forms a cake.

The Singles Super is left in store for a period of 2 – 3 weeks to allow the product to mature thus ensuring maximum conversion of the phosphorous to the water-soluble form.

7.2.2.8 Granulator Plants

The purpose of a granulator plant is to produce granules of N, NP and NPK fertiliser for an agricultural market. The product from a granulation plant is homogeneous chemical mixture of various chemical products. In some countries, fertiliser grades are usually denoted by the N (Nitrogen) to P (Phosphorus) to K (Potassium) ratios, and with the total percentage of plant food indicated in brackets thereafter, e.g. 2:3:2(22), which contains 22% NPK of which 2/(2 + 3 + 2) x 22 is N, i.e. 6.3%N. Similarly:

- ASN (mixture of NH_4NO_3 and $(NH_4)_2SO_4$ – 27%N
- MAP (mono Ammonium Phosphate) - 11%N, 22%P

There are two granulation plants at the production site of the SA Fertiliser Group for the production of complex granular N, NP and NPK fertilisers. The two plants are called Gran 1 and Gran 2 as evident in Fig 5-1 and the production processes comprise the following similar steps:

a) Raw Material Preparation:

For solid raw material preparation, the powder intermediates, KCl, $(NH_4)_2SO_4$, MAP and SSP are crushed, screened and put into different hoppers. Dolomitic Lime is also put inside a hopper, but via pneumatic conveyancing. The raw materials, required for granulation, are drawn from the various hoppers, in accordance with the required granulation recipe. The combined raw materials are then fed into the granulator or pug mill.

b) Granulation Loop

i) In Gran 2, the granulation loop consists of the pug mill, granulator, drier, screens and crushers. For products containing high levels of Ammonium Nitrate, the pug mill is used for initial granulation. For other grades, only the granulator is used. In the granulator itself, the liquid raw materials, such as Ammonium Nitrate Solution and pipe reactor products, are combined with the solid raw material as well as recycle from the plant. Steam and water are also occasionally introduced to aid granulation. In the pipe reactor, a mixture of ammonium phosphoric acid and water is reacted together to produce a saturated solution of fertiliser salts, of which the main reaction is:

$$H_3PO_4 + NH_3 \rightarrow NH_4(H_2PO_4)$$

ii) In Gran 3, sulphuric acid and scrubbing liquid are also fed into the pipe reactor. Gran 3 also has a pipe reactor, equipped in the dryer that is used during DAP production. The product, ex-granulator, is then dried in a rotary hot drier and then passed over the screens to separate product size particles from fines. Oversize product is crushed before being recycled through the granulator.

c) Cooling

Product size granules are cooled down in a fluidised bed cooler after which they are given a coating of oil and kaolin dust to enhance their storage properties

7.2.3 Description of Supply Chain Problem / Requirement

Optimisation calculations need to determine the best suite of NPK fertiliser packages, e.g. 1:2:2 (38), 1:2:3 (33), 16:1:3 (44) where *x:y:z* is the NPK ratio and (*ab*) is the mass fraction of NPK-based components, that are to be produced and distributed to maximise the satisfaction of crop/customer demand in South Africa. The supply of NPK fertiliser must be in accordance with crop/customer demand statistics published by the United Nations and the Fertiliser Society of Southern Africa, FSSA, FAO-UN, (2005), and also taking into account the fuzzy market demand of NPK fertiliser as well as abiding by an environmental requirement to minimise the discharge of hazardous chemicals.

7.2.4 Illustrated Diagram of Supply Chain Network

Fig 7-3 overleaf represents a comprehensive flow diagram of the entire fertiliser production and distribution network of the SA Fertiliser Group.

Figure 7-3: Comprehensive flow diagram of the entire fertiliser production and distribution network of the SA Fertiliser Group →

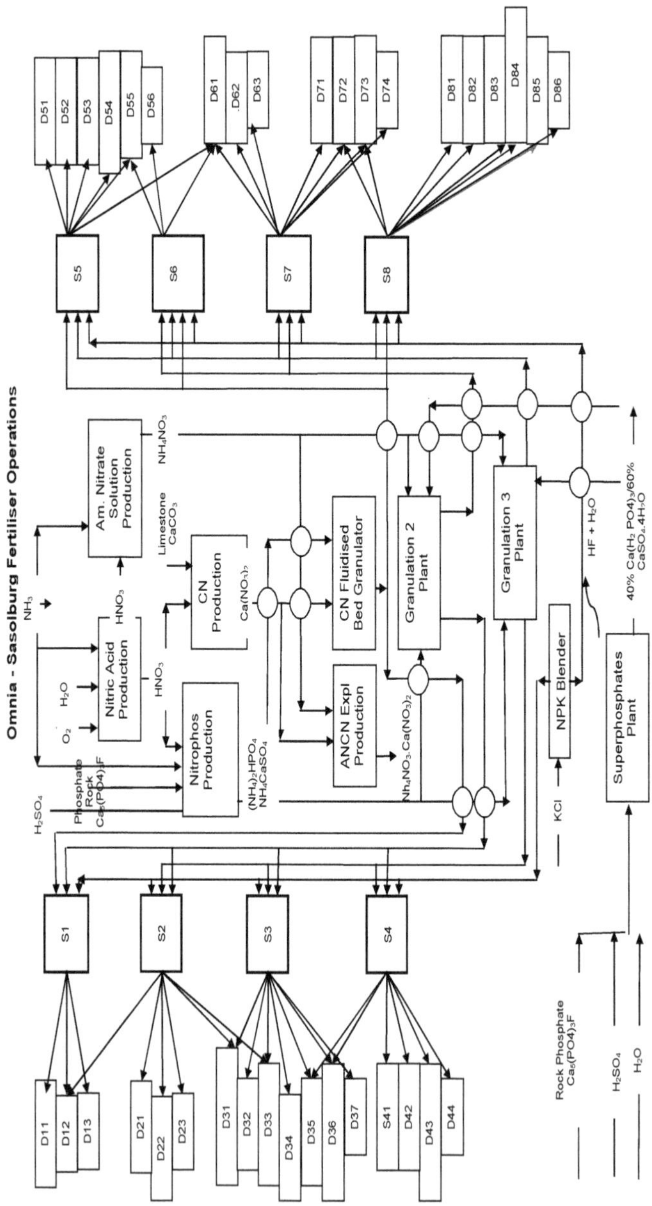
Omnia - Sasolburg Fertiliser Operations
Am. Nitrate Solution Production
Nitric Acid Production
CN Production
Nitrophos Production
CN Fluidised Bed Granulator
ANCN Expl Production
Granulation 2 Plant
Granulation 3 Plant
NPK Blender
Superphosphates Plant
S1
S2
S3
S4
S5
S6
S7
S8
D11
D12
D13
D21
D22
D23
D31
D32
D33
D34
D35
D36
D37
S41
D42
D43
D44
D51
D52
D53
D54
D55
D56
D61
D62
D63
D71
D72
D73
D74
D81
D82
D83
D84
D85
D86
Rock Phosphate
KCl

Such comprehensive process flow diagram, Fig. 7-3 is simplified and presented in two fundamental, constituent phases in Fig. 7-4, i.e. Production and Distribution, with the corresponding process nomenclature defined in Table 7-1

Table 7-1 Nomenclature Requirements

Category	Description
Indices/Sets	
Raw Materials	Ammonia (*A*), Sulphuric Acid (*S*), Rock phosphate (*RP*), Oxygen (*O*), Water (*W*), KCl (*K*), Limestone (*L*)
Intermediate Products	, e.g. Nitric Acid (*NA*), Nitrophos (*NP*), Superphosphates (*SP*), Ammonium Nitrate (*AN*), Calcium Nitrate (*CN*)
Final fertiliser products,	$(FBG)_i$, $(GR2)_j$, $(GR3)_k$ where *i, j, k* refer to the many various blends of NPK fertiliser exiting from the Granulators
P	Plants, e.g. Fluidised Bed Granulator (FBG), Granulator 1 (GR1) Granulator 2 (GR2), Superphosphates Plant (SP), Nitric Acid Plant (NA), Ammonium Nitrate Plant (AN), Nitrophos Plant (NP), NPK Blender (NPK)
S_n	Sites ($n = 1,\ldots, 8$) e.g. S_1, S_2
Variable Parameters	*Description*
(A)	Flowrate of NH_3 to the Fertiliser Production complex (t/yr)
$(A)_{AN}$	Flowrate of NH_3 to Ammonium Nitrate (AN) plant (t/yr)
$(A)_{NA}$	Flowrate of NH_3 to Nitric Acid (NA) plant (t/yr)
$(A)_{NP}$	Flowrate of NH_3 to Nitrophos (NP) plant (t/yr)
(S)	Total Flow'. of 98% Sulphuric Acid to Fertiliser complex (t/yr)
$(S)_{NP}$	Flowrate of 98%Sulphuric acid to NP plant (t/yr)
(RP)	Total Flowrate of Rock Phosphate to Fertiliser complex (t/yr)
$(RP)_{NP}$	Flowrate of Phosphate rock to NP plant (t/yr)
$(O)_{NA}$	Flowrate of Oxygen to NA plant (t/yr)
$(W)_{NA}$	Flowrate of Water to NA plant (t/yr)
$(NA)_{NA}$	Flowrate of Nitric Acid from NA plant (t/yr)
$(NA)_{AN}$	Flowrate of Nitric Acid from NA plant to AN plant (t/yr)
$(NA)_{NPI}$	Flowrate of Nitric Acid from NA plant to NP plant (t/yr)
$(NA)_{NPO}$	Flowrate of Nitric Acid from NP plant (t/yr)
$(Ca)_{NP}$	Flowrate of $Ca(NO_3)_2.4H_2O$ from NP plant (t/yr)
$(HF)_{NP}$	Discharge of Hydrogen Fluoride (HF) from NP plant (t/yr)
$(NA)_{CN}$	Flow of NA from NA plant to Calcium Nitrate (CN) plant (t/yr)

$(L)_{CN}$	Flowrate of Limestone ($CaCO_3$) to CN plant (t/yr)
$(NP)_{NP}$	Flowrate of Nitrophos from NP plant (t/yr)
$(CN)_{CN}$	Flowrate of 58%Calcium Nitrate from CN plant (t/yr)
$(CD)_{CN}$	Discharge of CO_2 from the CN plant (t/yr)
$(CN)_{ANCN}$	Flowrate of Calcium Nitrate to ANCN (explosives) plant (t/yr)
$(AN)_{ANCN}$	Flowrate of AN to ANCN (explosives) plant (t/yr)
$(ANCN)_{ANCN}$	Flowrate of ANCN from ANCN Plant (explosives) plant (t/yr)
$(NP)_{FBG}$	Flowrate of Nitrophos to Fluidised Bed Granulator (t/yr)
$(CN)_{FBG}$	Flowrate of Calcium Nitrate to Fluidised Bed Granulator (t/yr)
$(AN)_{ANCN}$	Flow of Ammonium Nitrate to ANCN (explosives) plant (t/yr)
$(FBG)_{FBG}$	Flowrate of granulated CN from FBG (t/yr)
$(NP)_{GR}$	Flowrate of Nitrophos to Granulator (t/yr)
$(AN)_{GR}$	Flowrate of Ammonium Nitrate to Granulator (t/yr)
$(SP)_{GR}$	Flowrate of Superphosphate to Granulator (t/day)
S_{SP}	Flowrate of Sulphuric Acid to Superphosphates plant (t/yr)
$(SP)_{SP}$	Flow of Superphosphate from Superphosphates plant (t/yr)
$(RP)_{SP}$	Flowrate of Phosphate Rock to Superphosphates plant (t/yr)
W_{SP}	Flowrate of Water to Superphosphates plant (t/yr)
$(NPK)_i^{Sn}$	Flowrate of *NPK* fertiliser to Site *n* (t/day) of NPK blend *i*
$(FBG)^{Sn}$	Flowrate of fluidised bed granular product to Site *n* (t/yr)
K_{GR}	Flow of *KCl* to Granulator
$(NPK)^i_{S1}$	Flow of NPK fert. of blend *i*, from Site 1 to market
$(NPK)^i_{S2}$	Flow of NPK fert. of blend *i*, from Site 2 to market
$(NPK)^i_{S3}$	Flow of NPK fert. of blend *i*, from Site 3 to market
$(NPK)_i^{S4Dm}$	Flow of NPK fert. of blend *i*, from Site 4 to depot *m*, (m = 1, 4)
$(NPK)_i^{S5Dm}$	Flow of NPK fert. of blend *i*, from Site 5 to depot *m*, (m = 1, 6)
$(NPK)_i^{S6Dm}$	Flow of NPK fert. of blend *i*, from Site 6 to depot *m*, (m = 1, 3)
$(NPK)_i^{S7Dm}$	Flow of NPK fert. of blend *i*, from Site 7 to depot *m*, (m = 1, 4)
$(NPK)_i^{S8Dm}$	Flow of NPK fert. of blend *i*, from Site 8 to depot *m*, (m = 1, 5)
$(FBG)^{S1Dm}$	Flow of FBG fert. from Site 1 to depot *m*, (m = 1, 3)
$(FBG)^{S2Dm}$	Flow of FBG fert. from Site 2 to depot *m*, (m = 1, 3)
$(FBG)^{S3Dm}$	Flow of FBG fert. from Site 3 to depot *m*, (m = 1, 6)
$(FBG)^{S4Dm}$	Flow of FBG fert. from Site 4 to depot *m*, (m = 1, 4)
$(FBG)^{S5Dm}$	Flow of FBG fert. from Site 5 to depot *m*, (m = 1, 6)
$(FBG)^{S6Dm}$	Flow of FBG fert. from Site 6 to depot *m*, (m = 1, 3)
$(FBG)^{S7Dm}$	Flow of FBG fert. from Site 7 to depot *m*, (m = 1, 4)
$(FBG)^{S8Dm}$	Flow of FBG fert. from Site 8 to depot *m*, (m = 1, 5)
$[N]^i$	Application Rate of Nitrogen (kg/ha) for crop *i* in Fert. Area
$[P]^i$	Application Rate of Phosphor. (kg/ha) for crop *i* in Fert. Area
$[K]^i$	Application Rate of Potassium (kg/ha) for crop *i* in Fert. Area
Constant Parameters	***Description***
(LA)	Total Fertile Land Area involved = 4,241,000 ha

$(d)_{yr}$	Effective operational days per annum = 320 (88% uptime)
$(NA)_{NA}$	Production flowrate of 59% Nitric Acid = 1200t/d
$(AN)_{AN}$	Flowrate of 89% Ammonium Nitrate from AN plant = 700t/d
$(ANCN)_{ANCN}$	2 x 75 ton batch reactors consisting of 52% CN & 48% AN. Each batch takes 6 hrs. to complete
$(FBG)_{FBG}$	Production capacity of granulated CN from FBG = 150 t/day, BOM = 78AN/140CN
$(GR)_{GR}$	Production capacity of Granulated Fertiliser from Granulator = 1,700 t/day
$(SP)_{SP}$	Production capacity of Superphosphates plant = 450 t/day for SSP and 150 t/day for MCP

Figure 7-4: Production and Distribution Flowchart

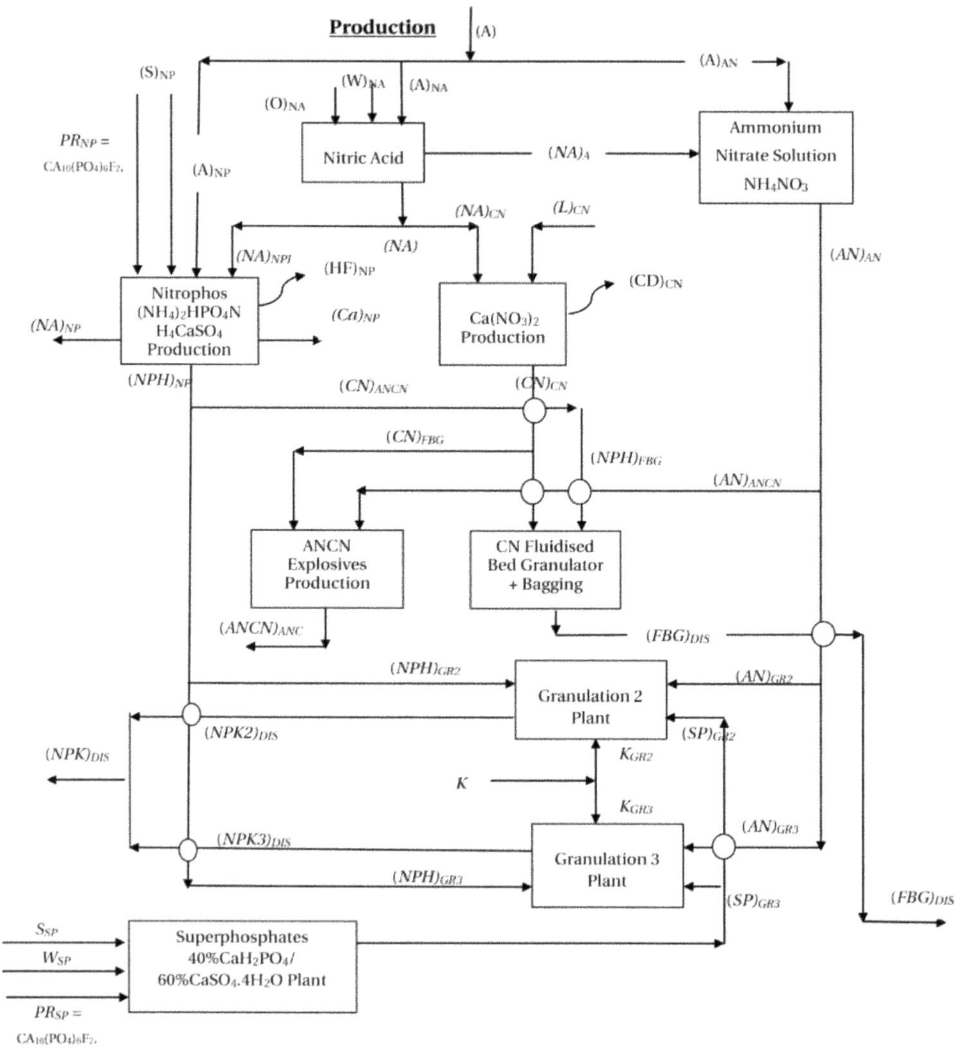

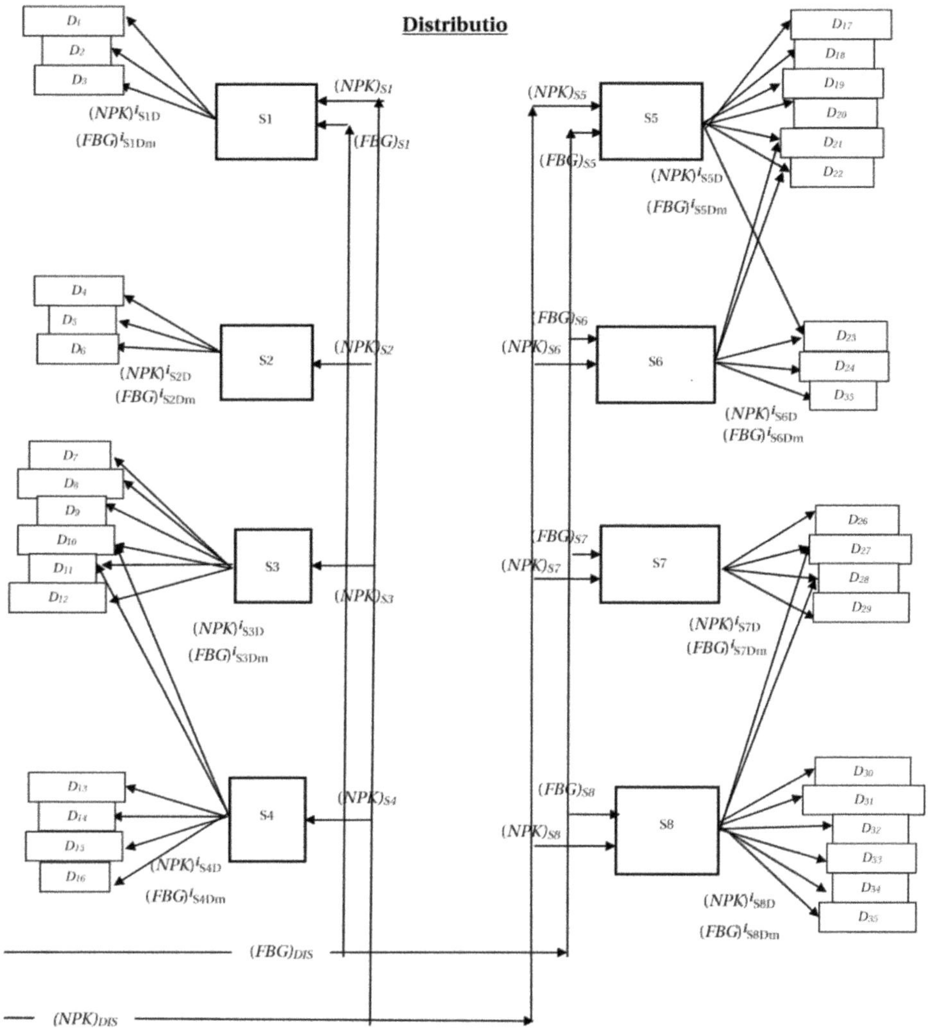
Distributio
S1
S2
S3
S4
S5
S6
S7
S8
(NPK)S1
(FBG)S1
(NPK)S2
(NPK)S3
(NPK)S4
(NPK)S5
(FBG)S5
(FBG)S6
(NPK)S6
(FBG)S7
(NPK)S7
(FBG)S8
(NPK)S8
(NPK)iS1D
(FBG)iS1Dm
(NPK)iS2D
(FBG)iS2Dm
(NPK)iS3D
(FBG)iS3Dm
(NPK)iS4D
(FBG)iS4Dm
(NPK)iS5D
(FBG)iS5Dm
(NPK)iS6D
(FBG)iS6Dm
(NPK)iS7D
(FBG)iS7Dm
(NPK)iS8D
(FBG)iS8Dm
(FBG)DIS
(NPK)DIS

7.2.5 Specify nature of Operational Uncertainty

Two forms of operational uncertainty are evident in this case study, and these are:

a) Multi-objective uncertainty evidenced by the need to maximise the production and distribution of an entire range of NPK fertiliser. In addition to this, there is also a need to minimise the discharge of hazardous chemicals whilst simultaneously maintaining the maximum production and distribution of NPK fertilisers

b) Fuzzy market demand uncertainty of NPK fertiliser

7.2.5.1 Multi-Objective Uncertainty

The situation of multi-objectivity arises because there is a requirement to maximise the production and distribution of the entire range of NPK fertiliser, i.e. Max $[\Sigma(NPK)^{i}]$ whilst simultaneously minimising the generation and discharge of hazardous effluent, carbon dioxide (CO_2) and hydrogen fluoride (HF), i.e. Min$[(HF)_{NP} + (CD)_{CN}]$ from the production process.

However, the chief requirement or objective of the SA Fertiliser Group is to maximise the production and distribution of multiple blends (15) of $(NPK)^{i}$ fertiliser in accordance with market demand, with i being representative of crop type, in accordance with market demand in South Africa. Such a requirement can be achieved by determining Max $[\Sigma(NPK)^{i}]$.

Therefore, there is a requirement to maximise the production and distribution of NPK fertilisers whilst simultaneously minimising the generation of hazardous effluent such as Hydrogen Fluoride (HF) from the Nitrophosphate unit plus Carbon Dioxide (CO_2) from the Calcium Nitrate unit. Such dual requirement is called a multi-objective requirement. Further to the research works of both Guillen and Grossmann (2009) and Liu and Papageorgiou (2013), the method of handling potentially conflicting multiple objectives, be they either minima or maxima, or a combination of the two, objectives, is by application of the ε-constraint method. This, of course, would provide a

Pareto optimal solution thereto, which means it would be impossible to make any one objective better off without making any other objective worse off. Typically, the principal objective of a project is selected as the objective of the MILP, whilst the others are allocated as ε-constraints with the ε-values determined by lower (min) or upper (max) bound limits

7.2.5.2 Fuzzy Demand Uncertainty

The nature of national, South African, demand for NPK-based fertiliser can best be gauged by viewing Table 7-2 in conjunction with Fig 7-5.

Table 7-2, published by the Fertiliser Society of South Africa (FSSA), shows the various crop-types in SA with their actual corresponding NPK fertiliser consumption requirements, specified in terms of the various N, P (P_2O_5) and K (K_2O) constituent figures (kg/ha) per associated crop-type land area involved (ha) for the year 2005.

Table 7-2: Crops fertilised and average rates of NPK fertilizer use in SA

Crop/Groups	Area	Rates, kg/ha of fertilised area				
	'000 ha	N	P2O5	P = 0.436P2O5	K2O	K = 0.83K2O
Field Crops						
Maize	3,354	55	30	13.08	6	4.98
Wheat	846	30	40	17.44	4	3.32
Sunflower	640	15	21	9.16	2	1.66
Skybeans	119	7	25	10.90	8	6.64
Sugar cane	432	92	57	24.85	133	110.39
2Lucerne	180	15	59	25.72	24	19.92
Other Pastures	395	50	44	19.18	7	5.81
Industrial Crops						
Tobacco	17	38	144	62.78	98	81.34
Cotton	42	36	22	9.59	3	2.49
Horticulture/Fruit						
Citrus	64	80	35	15.26	60	49.80
Subtropical fruits/nuts	50	180	57	24.85	240	199.20
Vines	180	50	36	15.70	24	19.92
Deciduous Fruit	50	110	159	69.32	83	68.89
Vegetables	95	170	159	69.32	120	99.60
Potatoes	50	170	160	69.76	120	99.60

FSSA' 2005

Fig 7-5, on the other hand, graphically demonstrates actual NPK fertiliser sales, from the SA Fertiliser Group – the predominant NPK fertiliser supplier in the country, between the years 2006 and 2009.

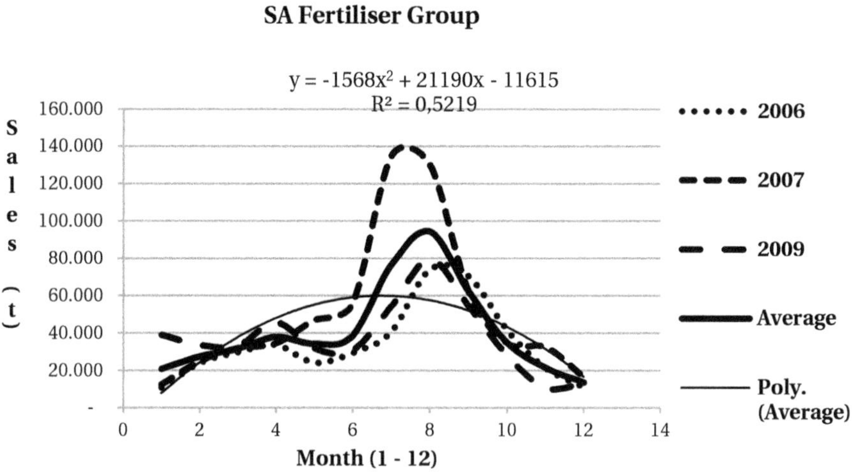

Figure 7-5: Actual fertiliser sales from the SA Fertiliser Group

Actual NPK fertiliser demand variety in SA, as clearly evident from Table 7-2, taken together with a representative national sample of actual NPK fertiliser sales in Fig 7-5, over a similar time horizon, clearly indicates that the nature of both NPK fertiliser market potential and fertiliser demand in South Africa is fuzzy, i.e. imprecise.

7.2.6 Specify Solution Approach

It was decided to break the overall problem down into three separate phases, i.e. Raw Material Production, Granulation and then Distribution, as these represent the three distinct processes involved in the overall production and distribution operations. The initial raw material production process, made up of a few continuous constituent production processes, e.g. nitric acid and nitrophosphate production, is a continuous operation dependent on market demand; followed by the key NPK fertiliser granulation process, which is a batch-continuous operation, even more intimately dependent on market demand. The final distribution phase is, to a large extent, dependent on market proximity.

Therefore in accordance with the 'Supply Chain Optimisation under prevailing conditions of Uncertainty' methodology, outlined in section 6, either a MILP (Mixed Integer Linear Program) or a MINLP (Mixed Integer Non-Linear Program) programming approach is preferred depending on the linear and/or non-linear nature of the objective function(s) and constraints.

Such fuzzy demand + MILP/MINLP programming approach would, according to Table 7-3 conform to that particular aspect of the originally derived methodology which is itemised in Table 6-1.

However, since there is also a requirement to maximise the production and distribution of NPK fertilisers whilst simultaneously minimising the generation of hazardous effluent, there is a multi-objective and, preferably, a Pareto optimum solution requirement. Therefore, according to the works of Guillen and Grossmann (2009) and Liu and Papageorgiou (2013), the method of handling potentially conflicting multiple objectives, be they either minima or maxima, or a combination of the two, objectives, is by application of the ε-constraint method, which would generate a *Pareto optimal* solution, Typically, the principal objective of a project, i.e. maximum production and distribution of NPK fertiliser, $\max\Sigma(NPK)^{i}$ is selected as the objective of the MILP, whilst the other(s), i.e. minimum effluent discharge, $\min[(HF)_{NP} + (CD)_{CN}]$, is assigned as the Pareto ε-constraint.

Table 7-3 Fuzzy Demand plus MILP/MINLP Programming Methodology

No.	Programming Step	Status
1	Detailed description of Supply Chain Network	Done
2	Description of Supply Chain problem/requirements	Done
3	Illustrated diagram of Supply Chain Network	Done
4	Define Nomenclature Requirements	Done
5	Specify nature of demand uncertainty and solution approach	Done
6	Define and formulate the MILP program - Define the Objective Function - Define the ε-constraint - Define all probabilistic/deterministic objectives and if the former, transform them into deterministic ones using the Lee and Olson (1985) method - Define all probabilistic/deterministic constraints and if the former, transform them into deterministic ones	 Done Done
7	Solve the MILP program	
8	Record the optimal results, i.e. Obj. Fn., ε-constraint and process variables	

7.2.7 Formulation of a MILP (Mixed Integer Linear Program)

Further to Table 7-3, the MILP for the NPK fertiliser production and distribution operations is as follows:

Objective Function: $\max \sum(NPK)^{i}$

Pareto ε-constraint: $(HF)_{NP} + (CD)_{CN} \geq \varepsilon$

- Production constraints
- Granulation constraints
- Distribution constraints

But before this can be done, it is first necessary to specify all operating parameters and assumptions.

Modeling Parameters and Assumptions

a) The SA Fertiliser Group NPK fertiliser optimisation initiative will be based upon the NPK fertiliser demands of different national crop types as published by the FSSA, 'Fertiliser use by crop type in SA Project Area' report, Table 7-4.

b) The requirement is to determine the optimum production and distribution of a suite (*multi-objective*) of NPK fertiliser blends, which may be efficiently and effectively mathematically represented by Max $[\Sigma(NPK)^i]$, in accordance with *fuzzy* SA crop fertiliser demand figures published by the FSSA. It is efficient since only one objective function is involved, and it is effective since the entire suite of NPK fertilisers will be maximised, automatically taking any interdependencies into account.

c) NPK fertiliser production and distribution figures must be on less fluctuating annual basis, which is also consistent with the annually reported FSSA NPK consumption figures

d) Model most be at NPK^i/crop level, $i = 1,.., 16$

e) Production utilises continuous processes up until the granulators and then batch thereafter

f) Raw material ratios per NPK blend can be determined from FSSA published NPK figures

g) For the purposes of this exercise, no fertiliser stock-holding is assumed, i.e. product manufactured goes directly to sales

h) Since the precise provincial distribution of crop types in SA is not essential to this study, they were estimated from FSSA graphical material, and further that provincial NPK fertiliser demand is shared equally between the sites and then equally again amongst the underlying depots

i) From a distribution perspective, there are eight (8) SA Fertiliser Group distribution sites situated strategically, from a market demand perspective, in the various SA provinces. For the purposes of this exercise, it is assumed that the sites in a particular represent, between them, the total fertiliser demand in that province. A number of strategically located depots (2 – 5) are associated with each site. Granular NPK product is distributed to these sites in accordance with customer orders, which are placed at those sites, and then transferred to the relevant depots for collection by the customer

j) It was decided, solely for the purposes of optimisation efficiency and practicality, to combine the two equal capacity (850t) granulators into one with double the capacity, 1700 t. This does not affect the optimisation process modeling at all since the model is only concerned about quantitative, and not qualitative, considerations

The formulation of an MILP program involves the definition of an Objective Function as well as the specification of all underlying constraints, both production and distribution related constraint

7.2.7.1 MILP Program Formulation

A] Objective Function

Optimum production (t) and distribution of a suite of NPK fertiliser in accordance with national FSSA,
reported different crop-type demand in SA
Max $(NPK)^{i}_{GR}$, i = NPK blend

B] Pareto -constraint

Further to the research work of (Liu and Papageorgiou (2013)), (Haimes and Chankong (1979)) and (Mavrotas and Florios (2013)), the range of an ε-constraint must lie between its possible bound values, and those would be between its highest possible bound, i.e. $[(HF)_{NP} + (CD)_{CN}]_{max} = 41{,}172$ t/yr and its lowest possible bound, i.e. $[(HF)_{NP} + (CD)_{CN}]_{min} = 38{,}880$ t/yr

Therefore -constraint: $[(HF)_{NP} + (CD)_{CN}] \geq 38{,}880$ (or $[(HF)_{NP} + (CD)_{CN}] \leq 41{,}172$)

Therefore -constraint: $[(HF)_{NP} + (CD)_{CN}] \geq\ = 38{,}880$

C] Production Model Constraints

Note: Molar Equivalence in a reaction: (with reference to Tables 1 & 2)	$n_i = x_i/Mw_i$; $x_i/x_j = n_iMw_i/n_jMw_j$; Mw_i= Molec. Wt. of *i* n_i = no. of moles of *i* $x_i = (n_i/n_j)(Mw_i/Mw_j)x_j$	n_i = no. moles *i*

a) Overall Balances

Overall: $(A) + (RP) + (S) + (O)_{NA} + (W) + (L)_{CN} + (K)_{GR} = (NPK)_{GR} + (FBG)_{DIS} + (ANCN)_{ANCN}$

Water: $(W) = (W)_{SP} + (W)_{NA} + (W)_{CN}$

Sulphur: $(S) = (S)_{SP} + (S)_{NP}$

Ammonia: $(A) = (A)_{AN} + (A)_{NA} + (A)_{NP}$

Phosphate Rock: $(RP) = (RP)_{NP} + (RP)_{SP}$

i) Nitric Acid Production

Production rate	$(NA)_{NA} \leq 1200$ t/d $\leq$ 384,000 t/yr
Strength of Nitric Acid produced	59%
Overall balance	$(O)_{NA} + (W)_{NA} + (A)_{NA} = (NA)_{NA}$
Chemical Processes: Oxidation	$5O_2 + 4NH_3 \quad 4NO + 6H_2O$
Absorption	$2O_2 + 4NO \quad 4NO_2$
Distillation	$4NO_2 + 6H_2O \quad 4HNO_3 + 2H_2 + 2H_2O$
Overall	$7O_2 + 4NH_3 \quad 4HNO_3 + 2H_2 + 2H_2O$
Molar equivalence:	$(A)_{NA} = (4/4)(17/63)(0.59)(NA)_{NA}$ $(A)_{NA} = 0.159(NA)_{NA}$
Additional water $(W)_{NA}$ required during distillation to produce 59% Nitric Acid	Water in final product - water produced in reaction
Water produced in reaction (applying (*W*) to (*NA*) molar equivalence)	$(n_W/n_{NA})(Mw_W/Mw_{NA})(0.59)(NA)_{NA}$ $= (6/4)(18/63)(0.59)(NA)_{NA}$ $= 0.253(NA)_{NA}$

Additional water required $= 0.41(NA)_{NA} - 0.253(NA)_{NA}$

Production Distribution $(NA)_{NA} = (NA)_{NPH} + (NA)_{CN} + (NA)_{AN}$

ii) Ammonium, Nitrate Production

Reaction $NH_3 + HNO_3 \quad NH_4NO_3 + H_2O$

Production rate $(AN)_{AN} \leq 700$ t/d $\leq 224{,}000$ t/yr

Strength of *AN* 89%

Overall balance $(A)_{AN} + (NA)_{AN} = (AN)_{AN} + (W)_{AN}$

Nitrogen balance $(14/17)(A)_{AN} + 0.59(14/63)(NA)_{NA} = 0.89(14/42)(AN)_{AN}$
or $0.7(A)_{AN} + 0.13(NA)_{AN} = 0.3(AN)_{AN}$

Water Vapour removed from process: $(W)_{AN} = (W)_{Reaction} - (1 - 0.89)(AN)_{AN}$

Applying molar equality $(W)_{Reaction} = (n_{H2O}/n_{NH4NO3})(Mw_{H2O}/Mw_{NH4NO3})(AN)_{AN} - 0.11(AN)_{AN}$
$= 10/42(AN)_{AN} = 0.238(AN)_{AN}$

Water removed $(W)_{AN} = 0.238(AN)_{AN} - 0.11(AN)_{AN}$
$(W)_{AN} = 0.128(AN)_{AN}$

Production Distribution $(AN)_{AN} = (AN)_{ANCN} + (AN)_{GR}$

iii) *CN* - Calcium Nitrate Production

Reaction $CaCO_3 + 2HNO_3 \quad Ca(NO_3)_2 + H_2O + CO_2$

Molar Equivalence $(CN)_{CN} = (0.59)(1/2)(164/63)(NA)_{CN}$
$(CN)_{CN} = 0.768(NA)_{CN}$

Release of water $(W)_{CN} = (1/1)(18/164)(CN)_{CN} = 0.11(CN)_{CN}$

Release of carbon dioxide $(CD)_{CN} = (1/1)(44/164)(CN)_{CN} = 0.268(CN)_{CN}$

Production rate $(CN)_{CN} \leq 500$ t/d $\leq 160{,}000$ t/yr

Strength of *CN* 58%

Overall balance $(L)_{CN} + (NA)_{CN} = (CN)_{CN} + (W)_{CN} + (CD)_{CN}$

Production Distribution $(CN)_{CN} = (CN)_{FBG} + (CN)_{ANCN}$

iv) Nitrophosphate Production

Reactions	1) $Ca_{10}(PO_4)_6F_2 + 20HNO_3 \rightarrow 6H_3PO_4 + 10Ca(NO_3)_2 + 2HF$ 2) $6H_3PO_4 + 10Ca(NO_3)_2 + 16H_2O \rightarrow 6H_3PO_4 + 4Ca(NO_3)_2.4H_2O + 6Ca(NO_3)_2$ 3) $6H_3PO_4 + 6Ca(NO_3)_2 + 20NH_3 + 7H_2SO_4 + 12H_2O$ $(NH_4)_2SO_4 + 6(NH4)H2PO4 + 12NH_4NO_3 + 6CaSO_4.2H_2O$
Effective overall Reaction	$Ca_{10}(PO_4)_6F_2 + 20HNO_3 + 28H_2O + 20NH_3 + 7H_2SO_4$ $[(NH_4)_2SO_4 + 6(NH_4)H_2PO_4 + 12NH_4NO_3 + 6CaSO_4.2H_2O] + 2HF + 4Ca(NO_3)_2.4H_2O$ $=$ Nitrophos + $2HF + 4Ca(NO_3)_2.4H_2O$
Production rate	$(NPH)_{NP} \leq 500$ t/d $\leq 160{,}000$ t/yr
Overall balance	$(S)_{NP} + (W)_{NP} + (A)_{NP} + (RP)_{NP} + (NA)_{NP} = (NPH)_{NP} + (HF)_{NP} + (Ca)_{NP}$

Molar Equivalence (molecular weights are in Table):

Use of 59% Nitric Acid, $(NA)_{NP}$	$(0.59)(NA)_{NP} = (n_{HNO3}/n_{RP})(Mw_{HNO3}/Mw_{RP})(RP)_{NP}$ $(NA)_{NP} = (20/1)(63/1008)(RP)_{NP}/0.59$ $(NA)_{NP} = 2.119(RP)_{NP}$
Similarly, for use of Ammonia $(A)_{NP}$	$(A)_{NP} = (20/1)(17/1008)(RP)_{NP}$ $(A)_{NP} = 0.337(RP)_{NP}$
Similarly, for use of Sulphuric Acid $(S)_{NP}$	$(S)_{NP} = (7/1)(98/1008)(RP)_{NP}$ $(S)_{NP} = 0.681(RP)_{NP}$
Similarly, for use of Water $(W)_{NP}$	$(W)_{NP} = (28/1)(18/1008)(RP)_{NP}$ $(W)_{NP} = 0.5(RP)_{NP}$
Similarly, for use of Nitrophos $[(NH_4)_2SO_4$ $(NH_4)H_2PO_4$, NH_4NO_3, $CaSO_4.2H_2O]$	$(NPH)_{NP} = (n_{NPH}/n_{RP})(Mw_{NPH}/Mw_{RP})(RP)_{NP}$ $(NPH)_{NP} = (1/1)(2814/1008)(RP)_{NP}$ $(NPH)_{NP} = 2.792(RP)_{NP}$
Hydrogen Fluoride $(HF)_{NP}$	$(HF)_{NP} = (2/1)(20/1008)(RP)_{NP}$ $(HF)_{NP} = 0.04(RP)_{NP}$
Production Distribution	$(NPH)_{NP} = (NPH)_{GR} + (NPH)_{FBG}$

v) ANCN Production

Reaction	$NH_4NO_3 + Ca(NO_3)_2$ $NH_4NO_3.Ca(NO_3)_2$
Production rate	$(ANCN)_{ANCN} \leq 25$t/hr $< 192{,}000$ t/yr
Overall balance	$(ANCN)_{ANCN} = (AN)_{ANCN} + (CN)_{ANCN}$

Nitrogen balance	$(56/244)(ANCN)_{ANCN} = (28/80)(AN)_{ANCN} + (28/164)(CN)_{ANCN}$ $0.23(ANCN)_{ANCN} = 0.35(AN)_{ANCN} + 0.17(CN)_{ANCN}$

vi) Fluidised Bed Granulator

Production rate	$(FBG)_{DIS} \leq 150t/d \leq 48{,}000\ t/yr$
Overall balance	$(NPH)_{FBG} + (CN)_{FBG} = (FBG)_{DIS}$
Ratio of $(NPH)_{FBG}$ to $CN)_{FBG}$ flow = 1:1	$(NPH)_{FBG} = (CN)_{FBG}$

vii) Superphosphates Production

Reactions	$Ca_{10}(PO_4)_6F_2 + 7H_2SO_4 + 3H_2O \quad 3Ca(H_2PO_4)_2H_2O + 2HF(g) + 7CaSO_4$ $3Ca(H_2PO_4)_2H_2O + 7CaSO_4 + 25H_2O \quad 3Ca(H_2PO_4)_2 + 7CaSO_4.4H_2O$
- overall reaction	$Ca_{10}(PO4)_6F_2 + 7H_2SO_4 + 28H_2O \quad 3Ca(H_2PO_4)_2 + 7CaSO_4.4H_2O + 2HF$
Production rate	$(SP)_{SP} \leq 500\ t/d \leq 160{,}000\ t/yr$
Overall balance	$(S)_{SP} + (RP)_{NP} + (W)_{SP} = (SP)_{SP}$

Phosphorous balance	$(186/1008)(RP)_{SP} = (186)/(2449)(SP)_{SP}$ or $0.076(SP)_{SP} = 0.185(RP)_{SP}$ or $(SP)_{SP} = 2.434(RP)_{SP}$
Sulphur balance	$(32/98)(S)_{SP} = (32/2449)(SP)_{SP}$ $0.327(S)_{SP} = 0.098(SP)_{SP}$ $(S)_{SP} = 0.013(SP)_{SP}$
Water Balance	$(W)_{SP} = (28)(18)/2449$ $(W)_{SP} = 0.206(SP)_{SP}$
Production Distribution	$(SP)_{SP} = (SP)_{GR}$

Table 7-4: Breakdown of Fert. Cons. per Crop Type over a Year in SA Project Area

Crop/Groups	Crop	Area $(LA)^i$	Rates, kg/ha of fertilised area				
	i	'000 ha	$[N]^i$	*P2O5*	$[P]^i$ = 0.436 P2O5	*K2O*	$[K]^i$ = 0.83 K2O
Field Crops							
Maize	1	2,384	55	30	13.08	6	4.98
Wheat	2	601	30	40	17.44	4	3.32
Sunflower	3	455	15	21	9.16	2	1.66
Skybeans	4	85	7	25	10.90	8	6.64
Sugar cane	5	307	92	57	24.85	133	110.39
Lucerne	6	128	15	59	25.72	24	19.92
Other Pastures	7	281	50	44	19.18	7	5.81
Mixed Crops							
Tobacco	8		38	144	62.78	98	81.34
Cotton	9	7	36	22	9.59	3	2.49
Vegetables	10	17	170	159	69.32	120	99.60
Potatoes	11	39	170	160	69.76	120	99.60
Horticulture/Fruit							
Citrus	12		80	35	15.26	60	49.80
Subtropical fruits/nuts	13	16	180	57	24.85	240	199.20
Vines	14	17	50	36	15.70	24	19.92
Deciduous Fruit	15		110	159	69.32	83	68.89

FSSA' 2005

Source: UN/FSSA - The NPK fertiliser consumption figures in Table 7-4 for the SA Fertiliser Group were compiled from the UN/FSSA magazine

Granulation Constraints (with reference to Table 7-4)

i) Overall Granulator equations with crop type *i* blend indication

$(NPK)^{i}_{GR} = (NPK)^{i}_{DIS}$

Overall (1 yr) Granulator Balances: $(NPH)_{GR} + (AN)_{GR} + (SP)_{GR} = (NPK)^{i}_{DIS} = (NPK)^{i}_{GR}$

Overall Production Limitation (i.e. 88% uptime): = $(NPK)^{i}_{GR} \leq (1700)320 \leq 544{,}000$ t/a

Overall (1 yr) Nitrophos Balances: $(NPH)_{GR} = (NPH)^{i}_{GR}$

Overall (1 yr) Ammonium Nitrate Balances:

$(AN)_{GR} = (AN)^{i}_{GR}$

Overall (1 yr) Potassium Chloride Balances: $(K)_{GR} = (K)^{i}_{GR}$

Overall (1 yr) Super Phosphate Balances: $(SP)_{GR} = (SP)^{i}_{GR}$

NPK Ratios: {NH_4NO_3, [$(NH_4)_2SO_4$ $(NH_4)H_2PO_4$, NH_4NO_3, $CaSO_4.2H_2O$],
' KCl, (36% $Ca(H_2PO_4)_2$ + 64% $CaSO_4.4H_2O$)}

ii) Granulation Parameters

Nitrophosphate = $(NH_4)_2SO_4.6(NH_4)H_2PO_4.12NH_4NO_3.6CaSO_4.2H_2O$

Molecular Weight (Mw) = (132 + 690 + 960 + 1032) = 2814

Mass percentage of N = 20(14)/2814 = 10%

Mass percentage of P = 6(31)/2814 = 6.61%

Superphosphate = $3Ca(H_2\,PO_4)_3$, $7CaSO_4.4H_2O$

Mw = 3(331) + 7(208) = 2449

Mass percentage P = 31)/2449 = 11.4%

Ammonium Nitrate = NH_4NO_3

Mw = 80 Mass Percentage N = 2(14)/80 = 35%

Potassium Chloride = KCl

Mw = 74.5 Mass Percentage K = 52.3%

iii) Granulator Batches - specific crop-type *NPK* fertilisers

Generic NPK fertiliser equations over the granulators per crop type, i:

- Overall $(NPH)^{i}_{GR} + (AN)^{i}_{GR} + (K)^{i}_{GR} + (SP)^{i}_{GR} = (NPK)^{i}_{GR}$
- Nitrogen (mass fraction N)$(NPH)^{i}_{GR}$ + (mass fraction N)$(AN)^{i}_{GR} \leq [N]^{i}(LA)^{i}$
  $0.1(NPH)^{i}_{GR} + 0.35(AN)^{i}_{GR} \leq [N]^{i}(LA)^{i}$
- Phosphorous (mass fraction P)$(NPH)^{i}_{GR}$ + (mass fraction P)$(SP)^{i}_{GR} \leq [P]^{i}(LA)^{i}$
  $0.0661(NPH)^{i}_{GR} + 0.114(SP)^{i}_{GR} \leq [P]^{i}(LA)^{i}$
- Potassium (mass fraction K)$(K)^{i}_{GR} \leq [K]^{i}(LA)^{i}$
  $0.523(K)^{i}_{GR} \leq [K]^{i}(LA)^{i}$

Maize (i = 1): Overall Balance $(NPH)^{1}_{GR} + (AN)^{1}_{GR} + (K)^{1}_{GR} + (SP)^{1}_{GR} = (NPK)^{1}_{GR}$

: Nitrogen Balance	$0.1(NPH)^1_{GR} + 0.35(AN)^1_{GR} \leq [N]^1(LA)^1$ $0.1(NPH)^1_{GR} + 0.35(AN)^1_{GR} \leq 55 \times 2384 \leq 131{,}120$
:Phosphorus Balance	$0.0661(NPH)^1_{GR} + 0.114(SP)^1_{GR} \leq [P]^1(LA)^1$ $0.0661(NPH)^1_{GR} + 0.114(SP)^1_{GR} \leq 13.08 \times 2384 \leq 31{,}183$
:Potassium Balance	$(K)^1_{GR} \leq 1.912(K)^1 \times (LA)^1$ $(K)^1_{GR} \leq 1.912 \times 2384 \times 4.98 = 22{,}700$
Wheat ($i = 2$): Overall Balance	$(NPH)^2_{GR} + (AN)^2_{GR} + (K)^2_{GR} + (SP)^2_{GR} = (NPK)^2_{GR}$
: Nitrogen Balance	$0.1(NPH)^2_{GR} + 0.35(AN)^2_{GR} \leq [N]^2(LA)^2$ $0.1(NPH)^2_{GR} + 0.35(AN)^2_{GR} \leq 30 \times 601 < 18{,}030$
:Phosphorus Balance	$0.0661(NPH)^2_{GR} + 0.114(SP)^2_{GR} \leq [P]^2(LA)^2$ $0.0661(NPH)^2_{GR} + 0.114(SP)^2_{GR} \leq 17.44 \times 601 \leq 10{,}481$
:Potassium Balance	$(K)^2_{GR} \leq 1.912[K]^i(LA)^i$ $(K)^2_{GR} \leq 1.912 \times 601 \times 3.32 \leq 3{,}815$
Sunflower ($i = 3$): Overall Balance	$(NPH)^3_{GR} + (AN)^3_{GR} + (K)^3_{GR} + (SP)^3_{GR} = (NPK)^3_{GR}$
: Nitrogen Balance	$0.1(NPH)^3_{GR} + 0.35(AN)^3_{GR} \leq [N]^3(LA)^3$ $0.1(NPH)^3_{GR} + 0.35(AN)^3_{GR2} \leq 15 \times 455 \leq 6{,}825$
:Phosphorus Balance	$0.0661(NPH)^3_{GR} + 0.114(SP)^3_{GR} \leq [P]^3(LA)^3$ $0.0661(NPH)^2_{GR} + 0.114(SP)^2_{GR} \leq 9.16 \times 455 \leq 4{,}168$
:Potassium Balance	$(K)^3_{GR} \leq 1.912[K]^3(LA)^3$ $(K)^3_{GR} \leq 1.912 \times 455 \times 1.66 \leq 1{,}444$
Skybeans ($i = 4$): Overall Balance	$(NPH)^4_{GR} + (AN)^4_{GR} + (K)^4_{GR} + (SP)^4_{GR} = (NPK)^4_{GR}$
: Nitrogen Balance	$0.1(NPH)^4_{GR} + 0.35(AN)^4_{GR} \leq [N]^4(LA)^4$ $0.1(\mathrm{NPH})^4_{GR} + 0.35(\mathrm{AN})^4_{GR} \leq 7 \times 85 \leq 595$
:Phosphorus Balance	$0.0661(NPH)^4_{GR} + 0.114(SP)^4_{GR} \leq [P]^4(LA)^4$ $0.0661(\mathrm{NPH})^4_{GR} + 0.114(\mathrm{SP})^4_{GR} \leq 10.9 \times 85 \leq 927$
:Potassium Balance	$(K)^4_{GR} \leq 1.912[K]^4(LA)^4$ $(K)^4_{GR} \leq 1.912 \times 85 \times 6.44 \leq 1{,}047$
Sugar cane ($i = 5$): Overall Balance	$(\mathrm{NPH})^5_{GR} + (\mathrm{AN})^5_{GR} + (\mathrm{K})^5_{GR} + (\mathrm{SP})^5_{GR} = (\mathrm{NPK})^5_{GR}$
: Nitrogen Balance	$0.1(NPH)^5_{GR} + 0.35(AN)^5_{GR} \leq [N]^5(LA)^5$ $0.1(\mathrm{NPH})^5_{GR} + 0.35(\mathrm{AN})^5_{GR} \leq 92 \times 307 \leq 28{,}244$

:Phosphorus Balance	$0.0661(NPH)^{5}_{GR} + 0.114(SP)^{5}_{GR} \leq [P]^{5}(LA)^{5}$
	$0.0661(NPH)^{5}_{GR} + 0.114(SP)^{5}_{GR} \leq 24.85 \text{ x } 307 \leq 7{,}629$
:Potassium Balance	$(K)^{5}_{GR} \leq 1.912[K]^{5}(LA)^{5}$
	$(K)^{5}_{GR} \leq 1.912 \text{ x } 307 \text{ x } 110.39 \leq 64{,}797$
Lucerne ($i = 6$): Overall Balance	$(NPH)^{6}_{GR} + (AN)^{6}_{GR} + (K)^{6}_{GR} + (SP)^{6}_{GR} = (NPK)^{6}_{GR}$
: Nitrogen Balance	$0.1(NPH)^{6}_{GR} + 0.35(AN)^{6}_{GR} \leq [N]^{6}(LA)^{6}$
	$0.1(NPH)^{6}_{GR} + 0.35(AN)^{6}_{GR} \leq 15 \text{ x } 128 < 1{,}920$
:Phosphorus Balance	$0.0661(NPH)^{6}_{GR} + 0.114(SP)^{6}_{GR} \leq [P]^{6}(LA)^{6}$
	$0.0661(NPH)^{6}_{GR} + 0.114(SP)^{6}_{GR} \leq 25.72 \text{ x } 128 \leq 3{,}292$
:Potassium Balance	$(K)^{6}_{GR} \leq 1.912[K]^{6}(LA)^{6}$
	$(K)^{6}_{GR} \leq 1.912 \text{ x } 128 \text{ x } 19.92 \leq 4{,}875$
Other Pastures ($i = 7$): Overall Balance	$(NPH)^{7}_{GR} + (AN)^{7}_{GR} + (K)^{7}_{GRj} + (SP)^{7}_{GR} = (NPK)^{7}_{GR}$
: Nitrogen Balance	$0.1(NPH)^{7}_{GR} + 0.35(AN)^{7}_{GR} \leq [N]^{7}(LA)^{7}$
	$0.1(NPH)^{7}_{GR} + 0.35(AN)^{7}_{GR} \leq 14{,}050$
:Phosphorus Balance	$0.0661(NPH)^{7}_{GR} + 0.114(SP)^{7}_{GR} \leq [P]^{7}(LA)^{7}$
	$0.0661(NPH)^{7}_{GR} + 0.114(SP)^{7}_{GR} \leq 19.18 \text{ x } 281 \leq 5{,}390$
:Potassium Balance	$(K)^{7}_{GR} \leq 1.912[K]^{7}(LA)^{7}$
	$(K)^{7}_{GR} \leq 1.912 \text{ x } 281 \text{ x } 5.81 < 3{,}122$
Cotton ($i = 9$): Overall Balance	$(NPH)^{9}_{GR} + (AN)^{9}_{GR} + (K)^{9}_{GR} + (SP)^{9}_{GRj} = (NPK)^{9}_{GR}$
: Nitrogen Balance	$0.1(NPH)^{9}_{GR} + 0.35(AN)^{9}_{GR} \leq [N]^{9}(LA)^{9}$
	$0.1(NPH)^{9}_{GR} + 0.35(AN)^{9}_{GR} \leq 36 \text{ x } 7 \leq 252$
:Phosphorus Balance	$0.0661(NPH)^{9}_{GR} + 0.114(SP)^{9}_{GR} \leq [P]^{9}(LA)^{9}$
	$0.0661(NPH)^{9}_{GR} + 0.114(SP)^{9}_{GR} \leq 9.59 \text{ x } 7 \leq 67$
:Potassium Balance	$(K)^{9}_{GR} \leq 1.912[K]^{9}(LA)^{9}$
	$(K)^{9}_{GR} \leq 1.912 \text{ x } 7 \text{ x } 2.49 < 33$
Vegetables ($i = 10$): Overall Balance	$(NPH)^{10}_{GR} + (AN)^{10}_{GR} + (K)^{10}_{GR} + (SP)^{10}_{GR} = (NPK)^{10}_{GR}$
: Nitrogen Balance	$0.1(NPH)^{10}_{GR} + 0.35(AN)^{10}_{GR} \leq [N]^{10}(LA)^{10}$
	$0.1(NPH)^{10}_{GR} + 0.35(AN)^{10}_{GR} \leq 170 \text{ x } 17 \leq 2{,}890$
:Phosphorus Balance	$0.0661(NPH)^{10}_{GR} + 0.114(SP)^{10}_{GR} \leq [P]^{10}(LA)^{10}$
	$0.0661(NPH)^{10}_{GR} + 0.114(SP)^{10}_{GR} \leq 69.32 \text{ x } 17 \leq 1{,}178$
:Potassium Balance	$(K)^{10}_{GR} \leq 1.912[K]^{10}(LA)^{10}$
	$(K)^{14}_{GR} \leq 1.912 \text{ x } 17 \text{ x } 99.6 \leq 3{,}327$

Potatoes ($i = 11$): Overall Balance	$(NPH)^{11}_{GR} + (AN)^{11}_{GR} + (K)^{11}_{GR} + (SP)^{11}_{GR} = (NPK)^{11}_{GR}$
: Nitrogen Balance	$0.1(NPH)^{11}_{GR} + 0.35(AN)^{11}_{GR} \leq [N]^{11}(LA)^{11}$ $0.1(NPH)^{11}_{GR} + 0.35(AN)^{11}_{GR} \leq 170 \times 39 \leq 6{,}630$
:Phosphorus Balance	$0.0661(NPH)^{11}_{GR} + 0.114(SP)^{11}_{GR} \leq [P]^{11}(LA)^{11}$ $0.0661(NPH)^{11}_{GR} + 0.114(SP)^{11}_{GR} \leq 69.76 \times 39 \leq 2{,}721$
:Potassium Balance	$(K)^{11}_{GR} \leq 1.912[K]^{15}(LA)^{15}$ $(K)^{11}_{GR} \leq 39 \times 99.6 < 3{,}884$
Subtropical fruits ($i = 13$): Overall Balance	$(NPH)^{13}_{GR} + (AN)^{13}_{GR} + (K)^{13}_{GR} + (SP)^{13}_{GR} = (NPK)^{13}_{GR}$
: Nitrogen Balance	$0.1(NPH)^{13}_{GR} + 0.35(AN)^{13}_{GR} \leq [N]^{13}(LA)^{13}$ $0.1(NPH)^{13}_{GR} + 0.35(AN)^{13}_{GR} \leq 2{,}880$
:Phosphorus Balance	$0.0661(NPH)^{13}_{GR} + 0.114(SP)^{13}_{GR} \leq [P]^{11}(LA)^{13}$ $0.0661(NPH)^{13}_{GR} + 0.114(SP)^{13}_{GR} \leq 24.85 \times 16 \leq 398$
:Potassium Balance	$(K)^{13}_{GR} \leq 1.912[K]^{13}(LA)^{13}$ $(K)^{13}_{GR} \leq 1.912 \times 16 \times 199.2 < 6{,}094$
Vines ($i = 14$): Overall Balance	$(NPH)^{14}_{GR} + (AN)^{14}_{GR} + (K)^{14}_{GR} + (SP)^{14}_{GR} = (NPK)^{14}_{GR}$
: Nitrogen Balance	$0.1(NPH)^{14}_{GR} + 0.35(AN)^{14}_{GR} \leq [N]^{14}(LA)^{14}$ $0.1(NPH)^{14}_{GR} + 0.35(AN)^{14}_{GR} \leq 50 \times 17 \leq 850$
:Phosphorus Balance	$0.0661(NPH)^{14}_{GR} + 0.114(SP)^{14}_{GR} \leq [P]^{14}(LA)^{14}$ $0.0661(NPH)^{14}_{GR} + 0.114(SP)^{14}_{GR} \leq 15.7 \times 17 \leq 267$
:Potassium Balance	$(K)^{14}_{GR} \leq 1.912[K]^{14}(LA)^{14}$ $(K)^{14}_{GR} \leq 1.912 \times 17 \times 19.92 \leq 647$

Now it follows from Table 7-4 that the various fertiliser crop type distribution limits can be compiled for the various distribution centres, $S_1 \ldots S_6$ of the SA Fertiliser Group and these are evident in Table 7-5

Distribution Constraints

(NPK)iGR = (NPK)iDIS
(NPK)iDIS = n=18 i=115(NPK)iSn
(FBG)DIS = n=18(FBG)Sn
Distribution Limits: (NPK)iSn < (p)iSn(LA)i(NPK)i/ha
(this relationship follows the work of Chen et al. (2007), Awudu & Zhang (2013) and Razmi et al. (2013) where, in this case, (p)iSn is the probability of demand for that particular crop-type in that particular area.

Table 7-5 Fertiliser crop-type distribution limits

Crop /Groups	Crop	Area	**Estimated Breakdown of Fertiliser Demand in section of SA (ha)**								**NPK[1] /ha**
	i	'000 ha	Region 1	Region 2				Region 3		Region 4	
Sites			S3	S2	S5	S6	S7	S4	S8	S1	
Field Crops											
Maize	1	2,384	177.09	295.15	295.15	295.15	295.15	238.93	238.93	548.69	284
Wheat	2	601	44.67	74.45	74.45	74.45	74.45	60.27	60.27	138.40	231
Sunflower	3	455	33.79	56.32	56.32	56.32	56.32	45.59	45.59	104.70	119
Skybeans	4	85	6.28	10.47	10.47	10.47	10.47	8.48	8.48	19.47	125
Sugar cane	5	307	22.81	38.02	38.02	38.02	38.02	30.77	30.77	70.67	881
Lucerne	6	128	9.50	15.84	15.84	15.84	15.84	12.82	12.82	29.45	306
Other Pastures	7	281	20.86	34.76	34.76	34.76	34.76	28.14	28.14	64.62	316
		4,241	315.00	525.00	525.00	525.00	525.00	425.00	425.00	976.00	
Mixed farming											
Tobacco	8	7.06	1.75	0.67	0.67	0.67	0.67	0.68	0.68	1.25	875
Cotton	9	17.44	4.33	1.66	1.66	1.66	1.66	1.69	1.69	3.08	191
Vegetables	10	39.44	9.81	3.76	3.76	3.76	3.76	3.81	3.81	6.97	1,403
Potatoes	11	20.76	5.16	1.98	1.98	1.98	1.98	2.01	2.01	3.67	1,406
		84.69	21.06	8.07	8.07	8.07	8.07	8.19	8.19	14.97	
Horticultur e/Fruit											
Citrus	12	16.00						8.00	8.00		547
fruits/nuts	13	17.00						8.50	8.50		1,491
Vines	14	5.00						2.50	2.50		344
Deciduous Fruit	15	38.00						19.00	19.00		1,100
		76.00						38.00	38.00		

FSSA'2005

Table 7-6: Distribution Limits by site

For the sake of this exercise, the probability $(p)^{i}_{Sn}$ of demand for fertiliser at all crops I at all sites n is assumed to be 1.

S1

$(NPK)^{1}_{S1} \leq (548.69)(284) \leq$ 155,828

$(NPK)^{2}_{S1} \leq (138.4)(231) \leq$ 31,970

$(NPK)^{3}_{S1} \leq (104.7)(119) \leq$ 12,459

$(NPK)^{4}_{S1} \leq (19.47)(125)$ $\leq$,438

$(NPK)^{5}_{S1} \leq (70.67)(881) \leq$ 62,260

$(NPK)^{6}_{S1} \leq (29.45)(306)$ $\leq$,012

$(NPK)^{7}_{S1} \leq (64.62)(316) \leq$ 31,975

$(NPK)^{8}_{S1} \leq (1.25)(875) \leq$ 1,094

$(NPK)^{9}_{S1} \leq (3.08)(191) \leq$ 1,977

$(NPK)^{10}_{S1} \leq (6.97)(1403) \leq$ 9,779

$(NPK)^{11}_{S1} \leq (3.67)(1406) \leq$ 5,160

S3

$(NPK)^{1}_{S3} \leq (295.15)(284) \leq$ 83,823

$(NPK)^{2}_{S3} \leq (74.45)(231) \leq$ 17,198

$(NPK)^{3}_{S3} \leq (56.32)(119) \leq$ 12,459

$(NPK)^{4}_{S3} \leq (10.47)(125) \leq$ 1,309

$(NPK)^{5}_{S3} \leq (38.02)(881) \leq$ 33,496

$(NPK)^{6}_{S3} \leq (15.84)(306) \leq$ 4,847

$(NPK)^{7}_{S3} \leq (34.76)(316) \leq$ 34,763

$(NPK)^{8}_{S3} \leq (0.67)(875) \leq 586$

$(NPK)^{9}_{S3} \leq (1.66)(191) \leq 317$

$(NPK)^{10}_{S3} \leq (3.76)(1403) \leq$ 5,275

$(NPK)^{11}_{S3} \leq (1.98)(1406) \leq$ 2,784

S5

$(NPK)^{1}_{S5} \leq (295.15)(284) \leq$ 83,823

$(NPK)^{2}_{S5} \leq (74.45)(231) \leq$ 17,198

$(NPK)^{3}_{S5} \leq (56.32)(119) \leq$ 12,459

$(NPK)^{4}_{S5} \leq (10.47)(125) \leq$ 1,309

$(NPK)^{5}_{S5} \leq (38.02)(881) \leq$ 33,496

$(NPK)^{6}_{S5} \leq (15.84)(306) \leq$ 4,847

$(NPK)^{7}_{S5} \leq (34.76)(316) \leq$ 34,763

$(NPK)^{8}_{S5} \leq (0.67)(875) \leq 586$

$(NPK)^{9}_{S5} \leq (1.66)(191) \leq 317$

$(NPK)^{10}_{S5} \leq (3.76)(1403) \leq$ 5,275

$(NPK)^{11}_{S5} \leq (1.98)(1406) \leq$ 2,784

S2

$(NPK)^{1}_{S2} \leq (295.15)(284) \leq$ 83,823

$(NPK)^{2}_{S2} \leq (74.45)(231) \leq$ 17,198

$(NPK)^{3}_{S2} \leq (56.32)(119) \leq$ 12,459

$(NPK)^{4}_{S2} \leq (10.47)(125)$ $\leq$,309

$(NPK)^{5}_{S2} \leq (38.02)(881) \leq$ 33,496

$(NPK)^{6}_{S2} \leq (15.84)(306)$ $\leq$,847

$(NPK)^{7}_{S2} \leq (34.76)(316) \leq$ 34,763

S4

$(NPK)^{1}_{S4} \leq (238.93)(284) \leq$ 67,8563

$(NPK)^{2}_{S4} \leq (60.27)(231) \leq$ 13,922

$(NPK)^{3}_{S4} \leq (45.59)(119) \leq$ 5,425

$(NPK)^{4}_{S4} \leq (8.48)(125) \leq$ 1,060

$(NPK)^{5}_{S4} \leq (30.77)(881) \leq$ 27,108

$(NPK)^{6}_{S4} \leq (12.82)(306) \leq$ 3,923

$(NPK)^{7}_{S4} \leq (28.14)(316) \leq$ 8,892

S6

$(NPK)^{1}_{S6} \leq (295.15)(284) \leq$ 83,823

$(NPK)^{2}_{S6} \leq (74.45)(231) \leq$ 17,198

$(NPK)^{3}_{S6} \leq (56.32)(119) \leq$ 12,459

$(NPK)^{4}_{S6} \leq (10.47)(125) \leq$ 1,309

$(NPK)^{5}_{S6} \leq (38.02)(881) \leq$ 33,496

$(NPK)^{6}_{S6} \leq (15.84)(306) \leq$ 4,847

$(NPK)^{7}_{S6} \leq (34.76)(316) \leq$ 34,763

$(NPK)^{8}_{S2} \leq (0.67)(875) \leq 586$	$(NPK)^{8}_{S4} \leq (0.68)(875) \leq 595$	$(NPK)^{8}_{S6} \leq (0.67)(875) \leq 586$
$(NPK)^{9}_{S2} \leq (1.66)(191) \leq 317$	$(NPK)^{9}_{S4} \leq (1.69)(191) \leq 323$	$(NPK)^{9}_{S6} \leq (1.66)(191) \leq 317$
$(NPK)^{10}_{S2} \leq (3.76)(1403) \leq 5{,}275$	$(NPK)^{10}_{S4} \leq (3.81)(1403) \leq 5{,}345$	$(NPK)^{10}_{S6} \leq (3.76)(1403) \leq 5{,}275$
$(NPK)^{11}_{S2} \leq (1.98)(1406) \leq 2{,}784$	$(NPK)^{11}_{S4} \leq (2.01)(1406) \leq 2{,}826$	$(NPK)^{11}_{S6} \leq (1.98)(1406) \leq 2{,}784$
	$(NPK)^{12}_{S4} \leq (8)(547) \leq 4{,}376$	
	$(NPK)^{13}_{S4} \leq (8.5)(1{,}491) \leq 12{,}674$	
	$(NPK)^{14}_{S4} \leq (2.5)(344) \leq 860$	
	$(NPK)^{15}_{S4} \leq (19)(1100) \leq 20{,}900$	

7.2.8 Specification of ε-Constraint MILP program

A MILP (Mixed Integer Linear Program), can now be structured that describes the production and distribution operations of the SA Fertiliser Group by integrating the various program elements of the production and distribution operations, the results of which are shown in Table 7-6

Table 7-7: ε-Constraint MILP (Mixed Integer Linear Program) Model for the production and distribution operations of the SA Fertiliser Group

SA Fertiliser Group – MILP (Mixed Integer Linear Program) Model

Production and Distribution Operations

OBJECTIVE FUNCTION

$\text{Max}(\ (NPK)^{i}_{GR}$

PARETO ε-CONSTRAINT

$(HF)_{NP} + (CD)_{CN} > 38{,}880$

CONSTRAINTS

Equality

$(A) + (S) + (O)_{NA} + (W) + (L)_{CN} + (RP) + (K)_{GR} = (NPK)_{GR} + (FBG)_{DIS} + (ANCN)_{ANCN}$

$(S) = (S)_{SP} + (S)_{NP}$

$(W) = (W)_{SP} + (W)_{NA} + (W)_{NP} + (W)_{CN}$

$(A) = (A)_{AN} + (A)_{NA} + (A)_{NP}$

$(RP) = (RP)_{NP} + (RP)_{SP}$

$(O)_{NA} + (W)_{NA} + (A)_{NA} = (NA)_{NA}$

$(A)_{NA} = 0.159(NA)_{NA}$

$(W)_{NA} = 0.157(NA)_{NA}$

$(NA)_{NA} = (NA)_{NPI} + (NA)_{CN} + (NA)_{AN}$

$(A)_{AN} + (NA)_{AN} = (AN)_{AN} + (W)_{AN}$

$0.7(A)_{AN} + 0.13(NA)_{AN} = 0.3(AN)_{AN}$

$(W)_{AN} = 0.128(AN)_{AN}$

$(AN)_{AN} = (AN)_{ANCN} + (AN)_{GR}$

$(L)_{CN} + (NA)_{CN} = (CN)_{CN} + (CD)_{CN} + (W)_{CN}$

$(CN)_{CN} = 0.768(NA)_{CN}$

$(L)_{CN} + (NA)_{CN} = (CN)_{CN} + (W)_{CN} + (CD)_{CN}$

$(CN)_{CN} = 0.768(NA)_{CN}$

$(W)_{CN} = 0.11(CN)_{CN}$

$(CD)_{CN} = 0.268(CN)_{CN}$

$(CN)_{CN} = (CN)_{FBG} + (CN)_{ANCN}$

$(AN)_{ANCN} + (CN)_{ANCN} = (ANCN)_{ANCN}$

$0.23(ANCN)_{ANCN} = 0.35(AN)_{ANCN} + 0.17(CN)_{ANCN}$

$(S)_{NP} + (W)_{NP} + (A)_{NP} + (RP)_{NP} + (NA)_{NP} = (NPH)_{NP} + (HF)_{NP} + (Ca)_{NP}$

$(NA)_{NP} = 2.119(RP)_{NP}$

$(A)_{NP} = 0.337(RP)_{NP}$

$(S)_{NP} = 0.681(RP)_{NP}$

$(W)_{NP} = 0.5(RP)_{NP}$

$(NPH)_{NP} = 2.792(RP)_{NP}$

$(HF)_{NP} = 0.038(RP)_{NP}$

$(NPH)_{NP} = (NPH)_{GR} + (NPH)_{FBG}$

$(NPH)_{NP} = (NPH)_{GR} + (NPH)_{FBG}$

$(NPH)_{FBG} + (CN)_{FBG} = (FBG)_{DIS}$

$(NPH)_{FBG} = (CN)_{FBG}$

$(S)_{SP} + (RP)_{SP} + (W)_{SP} = (SP)_{SP}$

$(SP)_{SP} = 2.434(RP)_{SP}$

$(S)_{SP} = 0.013(SP)_{SP}$

$(SP)_{SP} = (SP)_{GR}$

$(W)_{SP} = 0.206(SP)_{SP}$

$(AN)_{GR} = (AN)^{i}_{GR}$

$(NPH)_{GR} = (NPH)^{i}_{GR}$

$(K)_{GR} = (K)^{i}_{GR}$

$(SP)_{GR} = (SP)^{i}_{GR}$

$(NPH)^{1}_{GR} + (AN)^{1}_{GR} + (K)^{1}_{GR} + (SP)^{1}_{GR} = (NPK)^{1}_{GR}$

$(NPH)^{2}_{GR} + (AN)^{2}_{GR} + (K)^{2}_{GR} + (SP)^{2}_{GR} = (NPK)^{2}_{GR}$

$(NPH)^{3}_{GR} + (AN)^{3}_{GR} + (K)^{3}_{GR} + (SP)^{3}_{GR} = (NPK)^{3}_{GR}$

$(NPH)^{4}_{GR} + (AN)^{4}_{GR} + (K)^{4}_{GR} + (SP)^{4}_{GR} = (NPK)^{4}_{GR}$

$(NPH)^{5}_{GR} + (AN)^{5}_{GR} + (K)^{5}_{GR} + (SP)^{5}_{GR} = (NPK)^{5}_{GR}$

$(NPH)^{6}_{GR} + (AN)^{6}_{GR} + (K)^{6}_{GR} + (SP)^{6}_{GR} = (NPK)^{6}_{GR}$

$(NPH)^{7}_{GR} + (AN)^{7}_{GR} + (K)^{7}_{GRj} + (SP)^{7}_{GR} = (NPK)^{7}_{GR}$

$(NPH)^{9}_{GR} + (AN)^{9}_{GR} + (K)^{9}_{GR} + (SP)^{9}_{GRj} = (NPK)^{9}_{GR}$

$(NPH)^{11}_{GR} + (AN)^{11}_{GR} + (K)^{11}_{GR} + (SP)^{11}_{GR} = (NPK)^{11}_{GR}$

$(NPH)^{13}_{GR} + (AN)^{13}_{GR} + (K)^{13}_{GRj} + (SP)^{13}_{GR} = (NPK)^{13}_{GR}$

$(NPH)^{14}_{GR} + (AN)^{14}_{GR} + (K)^{14}_{GR} + (SP)^{14}_{GR} = (NPK)^{14}_{GR}$

$(NPK)^{i}_{GR} = {}_{n=1}^{8} {}_{i=1}^{15}(NPK)^{i}_{Sn}$

$(FBG)_{DIS} = {}_{n=1}^{8}(FBG)_{Sn}$

$(ANCN)_{ANCN} = 192{,}000$

$(FBG)_{DIS} = 48{,}000$

Inequality Constraints (<)

$(NA)_{NA} < 384{,}000$

$(AN)_{AN} < 224{,}000$

$(CN)_{CN} < 160{,}000$

$(NPH)_{NP} < 160{,}000$

$(SP)_{SP} < 160{,}000$

$(NPK)^{i}_{GR} < 544{,}000$

$0.1(NPH)^{1}_{GR} + 0.35(AN)^{1}_{GR} < 55 \times 2384 < 131{,}120$

$0.0661(NPH)^{1}_{GR} + 0.114(SP)^{1}_{GR} < 13.08 \times 2384 < 31{,}183$

$(K)^{1}_{GR} < 22{,}700$

$0.1(NPH)^{2}_{GR} + 0.35(AN)^{2}_{GR} < 18{,}030$

$0.0661(NPH)^{2}_{GR} + 0.114(SP)^{2}_{GR} < 10{,}481$

$(K)^{2}_{GR} < 3{,}815$

$0.1(NPH)^{3}_{GR} + 0.35(AN)^{3}_{GR2} < 6{,}825$

$0.0661(NPH)^{2}_{GR} + 0.114(SP)^{2}_{GR} < 4{,}168$

$(K)^{3}_{GR} < 1{,}444$

$0.1(NPH)^{4}_{GR} + 0.35(AN)^{4}_{GR} < 595$

$0.0661(NPH)^{4}_{GR} + 0.114(SP)^{4}_{GR} < 927$

$(K)^{4}_{GR} < 1{,}047$

$0.1(NPH)^{5}_{GR} + 0.35(AN)^{5}_{GR} < 28{,}244$

$0.0661(NPH)^{5}_{GR} + 0.114(SP)^{5}_{GR} < 7{,}629$

$(K)^{5}_{GR} < 64{,}797$

$0.1(NPH)^{6}_{GR} + 0.35(AN)^{6}_{GR} < 1{,}920$

$0.0661(NPH)^{6}_{GR} + 0.114(SP)^{6}_{GR} < 3{,}292$

$(K)^{6}_{GR} < 1.912 \times 128 \times 19.92 < 4{,}875$

$0.1(NPH)^{7}_{GR} + 0.35(AN)^{7}_{GR} < 14{,}050$

$0.0661(NPH)^{7}_{GR} + 0.114(SP)^{7}_{GR} < 5{,}390$

$(K)^{7}_{GR} < 3{,}122$

$0.1(NPH)^{9}_{GR} + 0.35(AN)^{9}_{GR} < 252$

$0.0661(NPH)^{9}_{GR} + 0.114(SP)^{9}_{GR} < 67$

$(K)^{9}_{GR} < 33$

$0.1(NPH)^{10}_{GR} + 0.35(AN)^{10}_{GR} < 2{,}890$

$0.0661(NPH)^{10}_{GR} + 0.114(SP)^{10}_{GR} < 1{,}178$

$(K)^{10}_{GR} < 3{,}327$

$0.1(NPH)^{11}_{GR} + 0.35(AN)^{11}_{GR} < 6{,}630$

$0.0661(NPH)^{11}_{GR} + 0.114(SP)^{11}_{GR} < 2{,}721$

$(K)^{11}_{GR} < 3{,}884$

$0.1(NPH)^{13}_{GR} + 0.35(AN)^{13}_{GR} < 2{,}880$

$0.0661(NPH)^{13}_{GR} + 0.114(SP)^{13}_{GR} < 398$

$(K)^{13}_{GR} < 6{,}094$

$0.1(NPH)^{14}_{GR} + 0.35(AN)^{14}_{GR} < 850$

$0.0661(NPH)^{14}_{GR} + 0.114(SP)^{14}_{GR} < 267$

$(K)^{14}_{GR} < 647$

$(NPK)^{1}_{S1} < 155{,}828$

$(NPK)^{2}_{S1} < 31{,}970$

$(NPK)^{3}_{S1} < 12{,}459$

$(NPK)^{4}_{S1} < 2{,}438$

$(NPK)^{5}_{S1} < 62{,}260$

$(NPK)^{6}_{S1} < < 9{,}012$

$(NPK)^{7}_{S1} < 31{,}975$

$(NPK)^{8}_{S1} < 1{,}094$

$(NPK)^{9}_{S1} < 31{,}977$

$(NPK)^{10}_{S1} < 9{,}779$

$(NPK)^{11}_{S1} < 5{,}160$

$(NPK)^{1}_{S2} < 83{,}823$

$(NPK)^{2}_{S2} < 17{,}198$

$(NPK)^{3}_{S2} < 12{,}459$

$(NPK)^{4}_{S2} < 1{,}309$

$(NPK)^{5}_{S2}) < 33{,}496$

$(NPK)^{6}_{S2} < 4{,}847$

$(NPK)^{7}_{S2} < 34{,}763$

$(NPK)^{8}_{S2} < 586$

$(NPK)^{9}_{S2} < 317$

$(NPK)^{10}_{S2} < 5{,}275$

$(NPK)^{11}_{S2} < 2{,}784$

$(NPK)^{1}_{S3} < 83{,}823$

$(NPK)^{2}_{S3} < 17{,}198$

$(NPK)^{3}_{S3} < 12{,}459$

$(NPK)^{4}_{S3} < 1{,}309$

$(NPK)^{5}_{S3} < 33{,}496$

$(NPK)^{6}_{S3} < 4{,}847$

$(NPK)^{7}_{S3} < 34{,}763$

$(NPK)^{8}_{S3} < 586$

$(NPK)^{9}_{S3} < 317$

$(NPK)^{10}_{S3} < 5{,}275$

$(NPK)^{11}_{S3} < 2{,}784$

$(NPK)^{1}_{S4} < 67{,}856$

$(NPK)^{2}_{S4} < 13{,}922$

$(NPK)^{3}_{S4} < 5{,}425$

$(NPK)^{4}_{S4}\ ) < 1{,}060$

$(NPK)^{5}_{S4} < 27{,}108$

$(NPK)^{6}_{S4}\ ) < 3{,}923$

$(NPK)^{7}_{S4} < 8{,}892$

$(NPK)^{8}_{S4}\ ) < 595$

$(NPK)^{9}_{S4} < 323$

$(NPK)^{10}_{S4} < 5{,}345$

$(NPK)^{11}_{S4} < 2{,}826$

$(NPK)^{12}_{S4} < 4{,}376$

$(NPK)^{13}_{S4} < 12{,}674$

$(NPK)^{14}_{S4} < 860$

$(NPK)^{15}_{S4} < 20{,}900$

$(NPK)^{1}_{S5}\ ) < 83{,}823$

$(NPK)^{2}_{S5} < 17{,}198$

$(NPK)^{3}_{S5}\ ) < 12{,}459$

$(NPK)^{4}_{S5} < 1{,}309$

$(NPK)^{5}_{S5}\ ) < 33{,}496$

$(NPK)^{6}_{S5} < 4{,}847$

$(NPK)^{7}_{S5} < 34{,}763$

$(NPK)^{8}_{S5} < 586$

$(NPK)^{9}_{S5} < 317$

$(NPK)^{10}_{S5} < 5{,}275$

$(NPK)^{11}_{S5} < 2{,}784$

$(NPK)^{1}_{S6} < 83{,}823$

$(NPK)^{2}_{S6} < 17{,}198$

$(NPK)^{3}_{S6} < 12{,}459$

$(NPK)^{4}_{S6} < 1{,}309$

$(NPK)^{5}_{S6} < 33{,}496$

$(NPK)^{6}_{S6} < 4{,}847$

$(NPK)^{7}_{S6} < 34{,}763$

$(NPK)^{8}_{S6} < 586$

$(NPK)^{9}_{S6} < 317$

$(NPK)^{10}_{S6} < 5{,}275$

$(NPK)^{11}_{S6} < 2{,}784$

$(NPK)^{1}_{S7} < 83{,}823$

$(NPK)^{2}_{S7} < 17{,}198$

$(NPK)^{3}_{S7} < 12{,}459$

$(NPK)^{4}_{S7} < 1{,}309$

$(NPK)^{5}_{S7} < 33{,}496$

$(NPK)^{6}_{S7} < 4{,}847$

$(NPK)^{7}_{S7} < 34{,}763$

$(NPK)^{8}_{S7} < 586$

$(NPK)^{9}_{S7} < 317$

$(NPK)^{10}_{S7} < 5{,}275$

$(NPK)^{11}_{S7} < 2{,}784$

$(NPK)^{1}_{S8} < 67{,}856$

$(NPK)^{2}_{S8} < 13{,}922$

$(NPK)^{3}_{S8} < 5{,}425$

$(NPK)^{4}_{S8} < 1{,}060$

$(NPK)^{5}_{S8} < 27{,}108$

$(NPK)^{6}_{S8} < 3{,}923$

$(NPK)^{7}_{S8} < 8{,}892$

$(NPK)^{9}_{S8} < 323$

$(NPK)^{10}{}_{S8} < 5{,}345$

$(NPK)^{11}{}_{S8} < 2{,}826$

$(NPK)^{13}{}_{S8} < 12{,}674$

$(NPK)^{14}{}_{S8} < 860$

Inequality Constraints (> 0)

$(A) > 0$

$(A)_{AN} > 0$

$(A)_{NA} > 0$

$(A)_{NP} > 0$

$(S) > 0$

$(S)_{NP} > 0$

$(RP) > 0$

$(RP)_{NP} > 0$

$(O)_{NA} > 0$

$(W)_{NA} > 0$

$(NA)_{NA} > 0$

$(NA)_{AN} > 0$

$(HF)_{NP} > 0$

$(Ca)_{NP} > 0$

$(W)_{NP} > 0$

$(NA)_{NP} > 0$

$(NA)_{CN} > 0$

$(AN)_{AN} > 0$

$(L)_{CN} > 0$

$(NPH)_{NP} > 0$

$(S)_{SP} > 0$

$(SP)_{SP} > 0$

$(RP)_{SP} > 0$

$(W)_{SP} > 0$

$(W)_{CN} > 0$

$(CO)_{CN} > 0$

$(CN)_{CN} > 0$

$(NPH)_{GR} > 0$

$(AN)_{GR} > 0$

$(CN)_{ANCN} > 0$

$(AN)_{ANCN} > 0$

$(ANCN)_{ANCN} > 0$

$(NPH)_{FBG} > 0$

$(CN)_{FBG} > 0$

$(SP)_{GR} > 0$

$(K)_{GR} > 0$

$(NPH)^{i}_{GR} > 0$

$(AN)^{i}_{GR} > 0$

$(SP)^{i}_{GR} > 0$

$(K)^{i}_{GR} > 0$

$(NPK)^{i}_{Sn} > 0,\ i = 1, 15;\ n = 1, 8$

$(FBG)_{Sn} > 0,\ n = 1.8$

7.2.9 Solution of ε-Constraint MILP program

The solution of this large MILP (Mixed Integer Linear Program), 243 program variables and 430 constraint equations required the use of an extended version of Microsoft Excel Solver ver. 2010, i.e. Premium Solver, the results from which are displayed in Table 7-7

Table 7-8: Solution to NPK fertiliser optimisation program under prevailing conditions of uncertainty

Pareto Optimal Solution

- Production and Distribution of NPK Fertiliser -

(Maximise total NPK Fertiliser Production)

Optimum overall production volume = 540,294 t/yr

Corresponding (ε-constraint) effluent discharge rate = 41,172 t/yr

RM Production		Granulation		Distribution			
(A)	103,882	$(NPK)^1_{GR}$	414,920	$(NPK)^1_{S1}$	-	$(NPK)^1_{S5}$	83,823
$(A)_{AN}$	23,514	$(NPK)^2_{GR}$	39,431	$(NPK)^2_{S1}$	-	$(NPK)^2_{S5}$	17,198
$(A)_{NA}$	61,056	$(NPK)^3_{GR}$	1,444	$(NPK)^3_{S1}$	6,630	$(NPK)^3_{S5}$	12,459
$(A)_{NP}$	19,312	$(NPK)^4_{GR}$	1,047	$(NPK)^4_{S1}$	2,438	$(NPK)^4_{S5}$	1,309
(S)	41,038	$(NPK)^5_{GR}$	64,797	$(NPK)^5_{S1}$	-	$(NPK)^5_{S5}$	33,496
$(S)_{NP}$	39,026	$(NPK)^6_{GR}$	4,875	$(NPK)^6_{S1}$	-	$(NPK)^6_{S5}$	4,847
(W)	136,674	$(NPK)^7_{GR}$	3,122	$(NPK)^7_{S1}$	2,880	$(NPK)^7_{S5}$	34,763
(RP)	120,913	$(NPK)^9_{GR}$	33	$(NPK)^9_{S1}$	398	$(NPK)^9_{S5}$	317
$(RP)_{NP}$	57,307	$(NPK)^{10}_{GR}$	-	$(NPK)^{10}_{S1}$	-	$(NPK)^{10}_{S5}$	5,275
$(O)_{NA}$	262,656	$(NPK)^{11}_{GR}$	3,884	$(NPK)^{11}_{S1}$	-	$(NPK)^{11}_{S5}$	2,784
$(W)_{NA}$	60,288	$(NPK)^{13}_{GR}$	6,094	$(NPK)^1_{S2}$	850	$(NPK)^1_{S6}$	83,823
$(NA)_{NA}$	384,000	$(NPK)^{14}_{GR}$	647	$(NPK)^2_{S2}$	259	$(NPK)^2_{S6}$	-
$(NA)_{AN}$	75,067			$(NPK)^3_{S2}$	-	$(NPK)^3_{S6}$	12,459
$(W)_{AN}$	11,187			$(NPK)^4_{S2}$	-	$(NPK)^4_{S6}$	1,281
$(HF)_{NP}$	2,292			$(NPK)^5_{S2}$	-	$(NPK)^5_{S6}$	-
$(Ca)_{NP}$	103,438			$(NPK)^6_{S2}$	-	$(NPK)^6_{S6}$	-
$(W)_{NP}$	28,653			$(NPK)^7_{S2}$	-	$(NPK)^7_{S6}$	-
$(NA)_{NP}$	121,433			$(NPK)^9_{S2}$	-	$(NPK)^9_{S6}$	-
$(NA)_{CN}$	187,500			$(NPK)^{10}_{S2}$	-	$(NPK)^{10}_{S6}$	-
$(AN)_{AN}$	87,395			$(NPK)^{11}_{S2}$	-	$(NPK)^{11}_{S6}$	-
$(L)_{CN}$	11,220			$(NPK)^1_{S3}$	-	$(NPK)^1_{S7}$	-
$(NPH)_{NP}$	160,000			$(NPK)^2_{S3}$	-	$(NPK)^2_{S7}$	-
$(S)_{SP}$	2,013			$(NPK)^3_{S3}$	-	$(NPK)^3_{S7}$	-

$(SP)_{SP}$	154,818		$(NPK)^{4}_{S3}$	1,309	$(NPK)^{4}_{S7}$	-
$(RP)_{SP}$	63,607		$(NPK)^{5}_{S3}$	33,496	$(NPK)^{5}_{S7}$	-
$(W)_{SP}$	31,893		$(NPK)^{6}_{S3}$	4,847	$(NPK)^{6}_{S7}$	-
$(W)_{CN}$	15,840		$(NPK)^{7}_{S3}$	34,763	$(NPK)^{7}_{S7}$	-
$(\mathrm{CD})_{CN}$	38,880		$(NPK)^{9}_{S3}$	317	$(NPK)^{9}_{S7}$	-
$(CN)_{CN}$	144,000		$(NPK)^{10}_{S3}$	5,275	$(NPK)^{10}_{S7}$	-
$(NPH)_{GR}$	266,171		$(NPK)^{11}_{S3}$	2,784	$(NPK)^{11}_{S7}$	-
$(AN)_{GR}$	15,395		$(NPK)^{1}_{S4}$	67,856	$(NPK)^{1}_{S8}$	-
$(CN)_{ANCN}$	120,000		$(NPK)^{2}_{S4}$	13,922	$(NPK)^{2}_{S8}$	-
$(AN)_{ANCN}$	72,000		$(NPK)^{3}_{S4}$	5,425	$(NPK)^{3}_{S8}$	-
$(ANCN)_{ANCN}$	192,000		$(NPK)^{4}_{S4}$	1,060	$(NPK)^{4}_{S8}$	-
$(NPH)_{FBG}$	24,000		$(NPK)^{5}_{S4}$	27,108	$(NPK)^{5}_{S8}$	-
$(CN)_{FBG}$	24,000		$(NPK)^{6}_{S4}$	3,923	$(NPK)^{6}_{S8}$	-
$(SP)_{GR}$	154,818		$(NPK)^{7}_{S4}$	8,892	$(NPK)^{7}_{S8}$	-
(K_{GR})	103,910		$(NPK)^{9}_{S4}$	323	$(NPK)^{9}_{S8}$	-
			$(NPK)^{10}_{S4}$	5,345	$(NPK)^{10}_{S8}$	-
			$(NPK)^{11}_{S4}$	2,826	$(NPK)^{11}_{S8}$	-
			$(NPK)^{13}_{S4}$	12,674	$(NPK)^{13}_{S3}$	-
			$(NPK)^{14}_{S4}$	860	$(NPK)^{14}_{S8}$	-
		540,294		246,460		293,834

7.2.10 Discussion of Results

The granular NPK fertiliser production and distribution operations case study represented an ideal platform for demonstrating the method and capability of the derived, 'Optimisation of a Supply Chain under Prevailing Combinational Conditions of Uncertainty' methodology. The chief reasons for this are that, firstly, all key aspects of an overall supply chain infrastructure, i.e. production, distribution and sales, were involved and that secondly, all prevailing (one) forms of operational uncertainty were accommodated, i.e. fuzziness, It must also be borne in mind that the planning aspect of multi-objectivity was also successfully realised in this case study in that the production and distribution of a range of NPK fertiliser was maximised, i.e. $\max\Sigma(NPK)^{i}$, whilst the production and discharge of hazardous effluent, i.e. $[(HF)_{NP} + (CD)_{CN}]$, was simultaneously minimised from the production process. The overall diversity and complexity of the problem was magnified by the existence of multiple (15) and similar (*NPK*) fertiliser product types.

As described in the solution procedure section of this problem example, linear program constraint modules were separately generated for the three key phases of the overall operation, i.e. production, granulation and distribution, and then these three were integrated together to create a comprehensive single constraint-based MILP for the entire overall operation. The creation of a program module for the production operation involved the sequential analysis and integration of all participating production units. Such an analysis included interface with the two NPK granulators but these were handled in a combined fashion on account of a single market dependence on the scale and variety of NPK fertilisers required by the marketplace. The program module, generated for the NPK fertiliser distribution and sales operations, was based on a detailed analysis of the (NPK) nutritional requirements, as well as the distribution, requirements of the various crop-types involved. Crop-type nutritional requirement data was obtained from an FSSA (Fertiliser Society of South Africa) report, FSSA'2005.

Once all the program constraints for the overall MILP were generated, it then became necessary to define the operational objectives for the overall operation to finalise the specification of the complete, overall MILP (objectives plus constraints). The principal operational objective was to <u>maximise</u> the production and distribution of NPK fertilisers in accordance with market demand but this was accompanied by an environmental objective to <u>minimise</u> the discharge of hazardous effluent, which in this case constituted Carbon Dioxide (CO_2) from the Calcium Nitrate (CN) unit and Hydrogen Fluoride (HF) from the Nitrophosphate (NP) unit. The incorporation of such consequent and ***planned*** <u>Multi-Objective</u> requirement involved an application of the Pareto ε-constraint method (Guillen and Grossmann (2009) and Liu and Papageorgiou (2013)) for solution whereby the one objective, i.e. minimal effluent discharge, was assigned as the ε-constraint.

The solution of this large MILP (Mixed Integer Linear Program), 243 program variables and 430 constraint equations, required the use of an extended version of Microsoft Excel Solver, and the optimum results from which are displayed in Table 7-7.

The key point that emerges from Table 7-7 is that the optimum overall production and associated distribution rate for a range of NPK fertilisers, in accordance with NPK fertiliser market demand, as described in Table 7-4, is 540,294 t/yr and the concomitant minimum total effluent discharge rate is 41,172 t/yr. This optimum production and distribution rate needs to be compared with the optimum NPK fertiliser annual production rate, optimised in isolation, to get a sense of the degree of optimality involved. Such optimum production rate, determined in isolation, is in fact exactly the same at 540,294 t/yr, i.e. a 100% degree of optimality achieved. Similarly, the calculated effluent discharge rate of 41,172 t/yr needs to be compared with the single objective, optimal minimum effluent flow MILP possible i.e. 38,880 t/yr to get a sense of the degree of optimality achieved in that area, i.e. [41,172 – (41,172 – 38,880)]/41,172 = 94.4%.

Also included in Table 7-7 are the optimum production, granulation and distribution figures for the entire range of NPK based fertilisers for the various different crop types. The Granulation figures give the total optimum annual NPK fertiliser production figures for each crop type irrespective of distribution destination and from a Distribution perspective, optimum distribution figures relate to NPK fertiliser market requirements at those particular sites (S_i), Such NPK fertiliser site demand figures were based on NPK unitary fertilisation requirement figures (kg/ hectare) and crop-type land areas as extracted from various identified tables in the FSSA 2005 report according to procedures described in Table 6-1.

The derived Planning and Optimisation of a Supply Chain under Uncertainty methodology was proven to be very effective in that 100% optimal effectiveness was achieved on NPK production and distribution throughput and that, further, 94.4% optimal efficiency was achieved on the overall effluent discharge rate.

7.3 Confirming Supply Chain Optimality

It now becomes necessary to confirm the optimality achieved with this application of the proposed, 'Planning and Optimisation Supply Chain, under uncertainty, Optimisation', methodology research effort. This is understandable common practise with most recent and modern day 'Supply Chain, under uncertainty, Planning, Management or Optimisation' research initiatives.

7.3.1 Selection and application of supply chain optimality confirmation technique

An analysis of the research literature revealed that a number of different methods can be employed to assess or confirm global supply chain optimality. These are itemised in Table 7-8 below and described thereunder:

Table 7-9: Assessing Supply Chain optimality

	Author(s)	**Research Work**
a)	Guillen & Grossman (2010)	Global optimality is confirmed by the novel spatial branch & bound method
b)	Liu & Papageorgiou (2013)	Dual comparative solution: ε-constraint method confirmed by lexicographic minmax method
c)	Haimes & Chankong (1979)	ε-constraint method confirmed by KKK (Karush-Kuhn-Tucker) sensitivity analysis
d)	Baky (2010)	Dual comparative solution: ε-constraint method confirmed by lexicographic minmax method
e)	Mavrotas & Florios (2013)	Use Augmecon2 to generate MILP Pareto optimal solutions that can be compared with either ε-constraint and/or lexicographic minmax method and/or by KKK sensitivity analysis
f)	Razmi et al (2013)	Augmented ε-constraint method may be applied for confirmation purposes
h)	Durango-Cohen & Sarutipand (2009)	Quadratic programming MILP framework that captures the relationship between demand and facility deterioration

a) Global optimality is confirmed by the novel spatial branch & bound method, introduced by Guillen & Grossman (2010) that guarantees optimality by exploiting the mathematical structure of the MINLP model.

b) Dual comparative solution whereby the ε-constraint MINLP solution method is confirmed by the lexicographic minmax solution method - Liu & Papageorgiou (2013). Further to these research work of Liu & Papageorgiou (2013), it was shown that a key disadvantage of the ε-constraint method, i.e. the selection of the critical ε-vector, could be overcome by application of the lexicographic minmax solution method, which could generate a fair solution by not showing preference to any objectives.

c) Haimes & Chankong (1979 demonstrated that MINLP optimality could be achieved by application of KKK sensitivity analysis. This, clearly, could be confirmed by application of something like the ε-constraint method, and vice versa.

d) Further to the dual comparative research work of Baky (2010), it was also shown that the ε-constraint method could be confirmed by the lexicographic minmax method

e) The Augmecon2 MILP solution technique, developed by Mavrotas & Florios (2013), was shown to generate MILP Pareto optimal solutions and that could be compared with either ε-constraint and/or lexicographic minmax method and/or by KKK sensitivity analysis

f) Razmi et al (2013) demonstrated that their augmented ε-constraint method, that involves a bi-objective stochastic optimisation solution formulation, may also be applied for MINLP solution optimality confirmation purposes.

g) The quadratic programming MILP framework, developed by Durango-Cohen & Sarutipand (2009) and that captures the relationship between demand and facility deterioration, could also be applied to confirm optimality

It therefore appears that the lexicographic minmax MILP optimisation technique, as indicated above, would be best suited to confirm MILP optimality, as determined by the ε-constraint method. However, further research into both the research works of Liu & Papageorgiou (2013) and Baky (2010) revealed that the Lexicographic minmax multi-objective optimisation technique is, in fact, a ML-MLOP mechanism, i.e. a multi-level, multi-objective MILP optimisation technique. This means that it can manage/optimise multiple performance objectives across multiple organisational, or hierarchical, levels as well as handling multiple performance objectives within each such level. Research also revealed that it is a fairly involved mechanism. Therefore, since this current research effort is only concerned with multiple objectives at one single management or one single organisational level, it would be far preferable, simpler and effective to identify such single level MILP optimisation technique.

Further research revealed that some original (single-level) multi-objective MILP optimisation research work had originally and successfully been done by Zimmerman (1978). This work had been refined by Bit et al. (1992) to create a multi-objective MILP optimisation technique that could provide for a non-dominated optimum solution, and which could then be regarded as an '*Optimum Compromise*' solution.

7.3.2 Application of selected optimality confirmation technique

The $\in$-constraint MILP that programmatically described the original agri-fertiliser production and distribution operations in Table 7-6 has been modified a trifle, in Table 7-9, to reflect the fact that the $\in$-constraint has now been replaced by an equivalent objective function, i.e. $(HF)_{NP} + (CD)_{CN} \geq 38{,}880$.

Table 7-10: Multi-Objective MILP (Mixed Integer Linear Program) Model for the production and distribution operations of the SA Fertiliser Group

SA Fertiliser Group – MILP (Mixed Integer Linear Program) Model

Production and Distribution Operations

OBJECTIVE FUNCTIONS

$\text{Max}(\ (NPK)^{i}_{GR}$

$(HF)_{NP} + (CD)_{CN} > 38{,}880$

CONSTRAINTS

Equality

$(A) + (S) + (O)_{NA} + (W) + (L)_{CN} + (RP) + (K)_{GR} = (NPK)_{GR} + (FBG)_{DIS} + (ANCN)_{ANCN}$

$(S) = (S)_{SP} + (S)_{NP}$

$(W) = (W)_{SP} + (W)_{NA} + (W)_{NP} + (W)_{CN}$

$(A) = (A)_{AN} + (A)_{NA} + (A)_{NP}$

$(RP) = (RP)_{NP} + (RP)_{SP}$

$(O)_{NA} + (W)_{NA} + (A)_{NA} = (NA)_{NA}$

$(A)_{NA} = 0.159(NA)_{NA}$

$(W)_{NA} = 0.157(NA)_{NA}$

$(NA)_{NA} = (NA)_{NPI} + (NA)_{CN} + (NA)_{AN}$

$(A)_{AN} + (NA)_{AN} = (AN)_{AN} + (W)_{AN}$

$0.7(A)_{AN} + 0.13(NA)_{AN} = 0.3(AN)_{AN}$

$(W)_{AN} = 0.128(AN)_{AN}$

$(AN)_{AN} = (AN)_{ANCN} + (AN)_{GR}$

$(L)_{CN} + (NA)_{CN} = (CN)_{CN} + (CD)_{CN} + (W)_{CN}$

$(CN)_{CN} = 0.768(NA)_{CN}$

$(L)_{CN} + (NA)_{CN} = (CN)_{CN} + (W)_{CN} + (CD)_{CN}$

$(CN)_{CN} = 0.768(NA)_{CN}$

$(W)_{CN} = 0.11(CN)_{CN}$

$(CD)_{CN} = 0.268(CN)_{CN}$

$(CN)_{CN} = (CN)_{FBG} + (CN)_{ANCN}$

$(AN)_{ANCN} + (CN)_{ANCN} = (ANCN)_{ANCN}$

$0.23(ANCN)_{ANCN} = 0.35(AN)_{ANCN} + 0.17(CN)_{ANCN}$

$(S)_{NP} + (W)_{NP} + (A)_{NP} + (RP)_{NP} + (NA)_{NP} = (NPH)_{NP} + (HF)_{NP} + (Ca)_{NP}$

$(NA)_{NP} = 2.119(RP)_{NP}$

$(A)_{NP} = 0.337(RP)_{NP}$

$(S)_{NP} = 0.681(RP)_{NP}$

$(W)_{NP} = 0.5(RP)_{NP}$

$(NPH)_{NP} = 2.792(RP)_{NP}$

$(HF)_{NP} = 0.038(RP)_{NP}$

$(NPH)_{NP} = (NPH)_{GR} + (NPH)_{FBG}$

$(NPH)_{NP} = (NPH)_{GR} + (NPH)_{FBG}$

$(NPH)_{FBG} + (CN)_{FBG} = (FBG)_{DIS}$

$(NPH)_{FBG} = (CN)_{FBG}$

$(S)_{SP} + (RP)_{SP} + (W)_{SP} = (SP)_{SP}$

$(SP)_{SP} = 2.434(RP)_{SP}$

$(S)_{SP} = 0.013(SP)_{SP}$

$(SP)_{SP} = (SP)_{GR}$

$(W)_{SP} = 0.206(SP)_{SP}$

$(AN)_{GR} = (AN)^{i}_{GR}$

$(NPH)_{GR} = (NPH)^{i}_{GR}$

$(K)_{GR} = (K)^{i}_{GR}$

$(SP)_{GR} = (SP)^{i}_{GR}$

$(NPH)^{1}_{GR} + (AN)^{1}_{GR} + (K)^{1}_{GR} + (SP)^{1}_{GR} = (NPK)^{1}_{GR}$

$(NPH)^{2}_{GR} + (AN)^{2}_{GR} + (K)^{2}_{GR} + (SP)^{2}_{GR} = (NPK)^{2}_{GR}$

$(NPH)^{3}_{GR} + (AN)^{3}_{GR} + (K)^{3}_{GR} + (SP)^{3}_{GR} = (NPK)^{3}_{GR}$

$(NPH)^{4}_{GR} + (AN)^{4}_{GR} + (K)^{4}_{GR} + (SP)^{4}_{GR} = (NPK)^{4}_{GR}$

$(NPH)^{5}_{GR} + (AN)^{5}_{GR} + (K)^{5}_{GR} + (SP)^{5}_{GR} = (NPK)^{5}_{GR}$

$(NPH)^{6}_{GR} + (AN)^{6}_{GR} + (K)^{6}_{GR} + (SP)^{6}_{GR} = (NPK)^{6}_{GR}$

$(NPH)^{7}_{GR} + (AN)^{7}_{GR} + (K)^{7}_{GRj} + (SP)^{7}_{GR} = (NPK)^{7}_{GR}$

$(NPH)^{9}{}_{GR} + (AN)^{9}{}_{GR} + (K)^{9}{}_{GR} + (SP)^{9}{}_{GRj} = (NPK)^{9}{}_{GR}$

$(NPH)^{11}{}_{GR} + (AN)^{11}{}_{GR} + (K)^{11}{}_{GR} + (SP)^{11}{}_{GR} = (NPK)^{11}{}_{GR}$

$(NPH)^{13}{}_{GR} + (AN)^{13}{}_{GR} + (K)^{13}{}_{GRj} + (SP)^{13}{}_{GR} = (NPK)^{13}{}_{GR}$

$(NPH)^{14}{}_{GR} + (AN)^{14}{}_{GR} + (K)^{14}{}_{GR} + (SP)^{14}{}_{GR} = (NPK)^{14}{}_{GR}$

$(NPK)^{i}{}_{GR} = {}_{n=1}{}^{8}\ {}_{i=1}{}^{15}(NPK)^{i}{}_{Sn}$

$(FBG)_{DIS} = {}_{n=1}{}^{8}(FBG)_{Sn}$

$(ANCN)_{ANCN} = 192{,}000$

$(FBG)_{DIS} = 48{,}000$

Inequality Constraints (<)

$(NA)_{NA} < 384{,}000$

$(AN)_{AN} < 224{,}000$

$(CN)_{CN} < 160{,}000$

$(NPH)_{NP} < 160{,}000$

$(SP)_{SP} < 160{,}000$

$(NPK)^{i}{}_{GR} < 544{,}000$

$0.1(NPH)^{1}{}_{GR} + 0.35(AN)^{1}{}_{GR} < 55 \times 2384 < 131{,}120$

$0.0661(NPH)^{1}{}_{GR} + 0.114(SP)^{1}{}_{GR} < 13.08 \times 2384 < 31{,}183$

$(K)^{1}{}_{GR} < 22{,}700$

$0.1(NPH)^{2}{}_{GR} + 0.35(AN)^{2}{}_{GR} < 18{,}030$

$0.0661(NPH)^{2}{}_{GR} + 0.114(SP)^{2}{}_{GR} < 10{,}481$

$(K)^{2}{}_{GR} < 3{,}815$

$0.1(NPH)^{3}{}_{GR} + 0.35(AN)^{3}{}_{GR2} < 6{,}825$

$0.0661(NPH)^{2}{}_{GR} + 0.114(SP)^{2}{}_{GR} < 4{,}168$

$(K)^{3}{}_{GR} < 1{,}444$

$0.1(NPH)^{4}{}_{GR} + 0.35(AN)^{4}{}_{GR} < 595$

$0.0661(NPH)^{4}{}_{GR} + 0.114(SP)^{4}{}_{GR} < 927$

$(K)^{4}{}_{GR} < 1{,}047$

$0.1(NPH)^{5}{}_{GR} + 0.35(AN)^{5}{}_{GR} < 28{,}244$

$0.0661(NPH)^{5}{}_{GR} + 0.114(SP)^{5}{}_{GR} < 7{,}629$

$(K)^{5}{}_{GR} < 64{,}797$

$0.1(NPH)^{6}{}_{GR} + 0.35(AN)^{6}{}_{GR} < 1{,}920$

$0.0661(NPH)^{6}{}_{GR} + 0.114(SP)^{6}{}_{GR} < 3{,}292$

$(K)^{6}{}_{GR} < 1.912 \times 128 \times 19.92 < 4{,}875$

$0.1(NPH)^{7}{}_{GR} + 0.35(AN)^{7}{}_{GR} < 14{,}050$

$0.0661(NPH)^{7}{}_{GR} + 0.114(SP)^{7}{}_{GR} < 5{,}390$

$(K)^{7}_{GR} < 3{,}122$

$0.1(NPH)^{9}_{GR} + 0.35(AN)^{9}_{GR} < 252$

$0.0661(NPH)^{9}_{GR} + 0.114(SP)^{9}_{GR} < 67$

$(K)^{9}_{GR} < 33$

$0.1(NPH)^{10}_{GR} + 0.35(AN)^{10}_{GR} < 2{,}890$

$0.0661(NPH)^{10}_{GR} + 0.114(SP)^{10}_{GR} < 1{,}178$

$(K)^{10}_{GR} < 3{,}327$

$0.1(NPH)^{11}_{GR} + 0.35(AN)^{11}_{GR} < 6{,}630$

$0.0661(NPH)^{11}_{GR} + 0.114(SP)^{11}_{GR} < 2{,}721$

$(K)^{11}_{GR} < 3{,}884$

$0.1(NPH)^{13}_{GR} + 0.35(AN)^{13}_{GR} < 2{,}880$

$0.0661(NPH)^{13}_{GR} + 0.114(SP)^{13}_{GR} < 398$

$(K)^{13}_{GR} < 6{,}094$

$0.1(NPH)^{14}_{GR} + 0.35(AN)^{14}_{GR} < 850$

$0.0661(NPH)^{14}_{GR} + 0.114(SP)^{14}_{GR} < 267$

$(K)^{14}_{GR} < 647$

$(NPK)^{1}_{S1} < 155{,}828$

$(NPK)^{2}_{S1} < 31{,}970$

$(NPK)^{3}_{S1} < 12{,}459$

$(NPK)^{4}_{S1} < 2{,}438$

$(NPK)^{5}_{S1} < 62{,}260$

$(NPK)^{6}_{S1} < \; < 9{,}012$

$(NPK)^{7}_{S1} < 31{,}975$

$(NPK)^{8}_{S1} < 1{,}094$

$(NPK)^{9}_{S1} < 31{,}977$

$(NPK)^{10}_{S1} < 9{,}779$

$(NPK)^{11}_{S1} < 5{,}160$

$(NPK)^{1}_{S2} < 83{,}823$

$(NPK)^{2}_{S2} < 17{,}198$

$(NPK)^{3}_{S2} < 12{,}459$

$(NPK)^{4}_{S2} < 1{,}309$

$(NPK)^{5}_{S2}) < 33{,}496$

$(NPK)^{6}_{S2} < 4{,}847$

$(NPK)^{7}_{S2} < 34{,}763$

$(NPK)^{8}_{S2} < 586$

$(NPK)^{9}_{S2} < 317$

$(NPK)^{10}_{S2} < 5{,}275$

$(NPK)^{11}_{S2} < 2{,}784$

$(NPK)^{1}_{S3} < 83{,}823$

$(NPK)^{2}_{S3} < 17{,}198$

$(NPK)^{3}_{S3} < 12{,}459$

$(NPK)^{4}_{S3} < 1{,}309$

$(NPK)^{5}_{S3} < 33{,}496$

$(NPK)^{6}_{S3} < 4{,}847$

$(NPK)^{7}_{S3} < 34{,}763$

$(NPK)^{8}_{S3} < 586$

$(NPK)^{9}_{S3} < 317$

$(NPK)^{10}_{S3} < 5{,}275$

$(NPK)^{11}_{S3} < 2{,}784$

$(NPK)^{1}_{S4} < 67{,}856$

$(NPK)^{2}_{S4} < 13{,}922$

$(NPK)^{3}_{S4} < 5{,}425$

$(NPK)^{4}_{S4}) < 1{,}060$

$(NPK)^{5}_{S4} < 27{,}108$

$(NPK)^{6}_{S4}) < 3{,}923$

$(NPK)^{7}_{S4} < 8{,}892$

$(NPK)^{8}_{S4}) < 595$

$(NPK)^{9}_{S4} < 323$

$(NPK)^{10}_{S4} < 5{,}345$

$(NPK)^{11}_{S4} < 2{,}826$

$(NPK)^{12}_{S4} < 4{,}376$

$(NPK)^{13}_{S4} < 12{,}674$

$(NPK)^{14}_{S4} < 860$

$(NPK)^{15}_{S4} < 20{,}900$

$(NPK)^{1}_{S5}) < 83{,}823$

$(NPK)^{2}_{S5} < 17{,}198$

$(NPK)^{3}_{S5}) < 12{,}459$

$(NPK)^{4}_{S5} < 1{,}309$

$(NPK)^{5}_{S5}) < 33{,}496$

$(NPK)^{6}_{S5} < 4{,}847$

$(NPK)^{7}{}_{S5} < 34{,}763$

$(NPK)^{8}{}_{S5} < 586$

$(NPK)^{9}{}_{S5} < 317$

$(NPK)^{10}{}_{S5} < 5{,}275$

$(NPK)^{11}{}_{S5} < 2{,}784$

$(NPK)^{1}{}_{S6} < 83{,}823$

$(NPK)^{2}{}_{S6} < 17{,}198$

$(NPK)^{3}{}_{S6} < 12{,}459$

$(NPK)^{4}{}_{S6} < 1{,}309$

$(NPK)^{5}{}_{S6} < 33{,}496$

$(NPK)^{6}{}_{S6} < 4{,}847$

$(NPK)^{7}{}_{S6} < 34{,}763$

$(NPK)^{8}{}_{S6} < 586$

$(NPK)^{9}{}_{S6} < 317$

$(NPK)^{10}{}_{S6} < 5{,}275$

$(NPK)^{11}{}_{S6} < 2{,}784$

$(NPK)^{1}{}_{S7} < 83{,}823$

$(NPK)^{2}{}_{S7} < 17{,}198$

$(NPK)^{3}{}_{S7} < 12{,}459$

$(NPK)^{4}{}_{S7} < 1{,}309$

$(NPK)^{5}{}_{S7} < 33{,}496$

$(NPK)^{6}{}_{S7} < 4{,}847$

$(NPK)^{7}{}_{S7} < 34{,}763$

$(NPK)^{8}{}_{S7} < 586$

$(NPK)^{9}{}_{S7} < 317$

$(NPK)^{10}{}_{S7} < 5{,}275$

$(NPK)^{11}{}_{S7} < 2{,}784$

$(NPK)^{1}{}_{S8} < 67{,}856$

$(NPK)^{2}{}_{S8} < 13{,}922$

$(NPK)^{3}{}_{S8} < 5{,}425$

$(NPK)^{4}{}_{S8} < 1{,}060$

$(NPK)^{5}{}_{S8} < 27{,}108$

$(NPK)^{6}{}_{S8} < 3{,}923$

$(NPK)^{7}{}_{S8} < 8{,}892$

$(NPK)^{9}{}_{S8} < 323$

$(NPK)^{10}{}_{S8} < 5,345$

$(NPK)^{11}{}_{S8} < 2,826$

$(NPK)^{13}{}_{S8} < 12,674$

$(NPK)^{14}{}_{S8} < 860$

Inequality Constraints (> 0)

$(A) > 0$

$(A)_{AN} > 0$

$(A)_{NA} > 0$

$(A)_{NP} > 0$

$(S) > 0$

$(S)_{NP} > 0$

$(RP) > 0$

$(RP)_{NP} > 0$

$(O)_{NA} > 0$

$(W)_{NA} > 0$

$(NA)_{NA} > 0$

$(NA)_{AN} > 0$

$(HF)_{NP} > 0$

$(Ca)_{NP} > 0$

$(W)_{NP} > 0$

$(NA)_{NP} > 0$

$(NA)_{CN} > 0$

$(AN)_{AN} > 0$

$(L)_{CN} > 0$

$(NPH)_{NP} > 0$

$(S)_{SP} > 0$

$(SP)_{SP} > 0$

$(RP)_{SP} > 0$

$(W)_{SP} > 0$

$(W)_{CN} > 0$

$(CO)_{CN} > 0$

$(CN)_{CN} > 0$

$(NPH)_{GR} > 0$

$(AN)_{GR} > 0$

$(CN)_{ANCN} > 0$

$(AN)_{ANCN} > 0$

$(ANCN)_{ANCN} > 0$

$(NPH)_{FBG} > 0$

$(CN)_{FBG} > 0$

$(SP)_{GR} > 0$

$(K)_{GR} > 0$

$(NPH)^{i}_{GR} > 0$

$(AN)^{i}_{GR} > 0$

$(SP)^{i}_{GR} > 0$

$(K)^{i}_{GR} > 0$

$(NPK)^{i}_{GR} > 0,\ i = 1, 15$

$(FBG)_{DIS} > 0$

$(NPK)^{i}_{Sn} > 0,\ i = 1, 15;\ n = 1, 8$

$(FBG)_{Sn} > 0,\ n = 1.8$

Further to the research work of Bit et al. (1992), the procedure for generating an optimal compromise solution for this multi-objective MILP is applied as in Table 7-11 below:

Table 7-11: Optimum Compromise Solution technique

Step 1: The multi-objective problem is solved as a single objective problem, using each objective in turn as the single objective function whilst using all the others as criteria.

Therefore Max $\Sigma(NPK)_{DIS} = 540{,}294$ t/yr

subject to constraints (1) – (238)

Similarly, Min$[(HF)_{NP} + (CD)_{CN}] = 38{,}880$ t/yr

subject to constraints (1) – (238)

Step 2: From the results of step 1, the corresponding decision-variable values are determined for each optimum objective.

Effluent discharge rate corresponding to $\Sigma(NPK)_{DIS}{}^{Max} = 41{,}172$ t/yr

Similarly, agri-fertiliser, (NPK) production and distribution rate corresponding to $[(HF)_{NP} + (CD)_{CN}]^{Min} = 55{,}776$ t/yr

Step 3: From step 2, a lower and upper bound-value is determined for the set of solutions for each objective function, Z_k. The lower bound is designated L_k, whilst the upper bound is designated U_k.

	U_k	L_k
Max$(NPK)_{DIS}$	540,294	55,776
Min(Effluent)	38,880	41,172

Step 4: Formulation of required membership functions, u_k (x)

$$u_k(x) = \begin{cases} 1 & ;\text{if } Z_k \leq L_k \\ 1 - (Z_k - L_k) / (U_k - L_k) & ;\text{if } L_k < Z_k < U_k \\ 0 & ;\text{if } Z_k \geq U_k \end{cases}$$

Therefore:

Objective 1: Max[$\Sigma(NPK)_{DIS}$	Objective 2: Min[$(HF)_{NP} + (CD)_{CN}$]
$U_1(x) = 1 - (Z_1 - L_1) / (/ (U_1 - L_1))$	$U_2(x) = 1 - (Z_2 - L_2) / (/ (U_2 - L_2))$
$U_1(x) = 1 - (Z_1 - 55{,}776) / (540{,}294 - 55{,}776)$	$U_2(x) = 1 - (Z_2 - 41{,}172) / (38{,}880 - 41{,}172)$
$U_1(x) = 1 - (Z_1 - 55{,}776) / (484{,}518)$	$U_2(x) = 1 - (Z_2 - 41{,}172) / (-2{,}292)$
$U_1(x) = 0.0000021\ Z_1 + 0.115$	$U_2(x) = 1 - (-0.00044 Z_2 + 17.963)$

Step 5: Maximise the membership functions, as per the same set of constraints, and, according to Bit et al. (1992), the highest membership function value corresponds to the optimum solution set, i.e.:

Max $[(u_1(x)] = 1.250$ and Max $[(u_2(x)] = 1.153$

Therefore Optimum $(u_1(x) = 1.250$ and the corresponding compromise optimal solution is set out in Table 7-12

Table 7-12: Compromise optimal solution set of NPK fertiliser production and distribution under uncertainty

Compromise Optimal Solution Set

- Production and Distribution of NPK Fertiliser -

(Maximise total NPK Fertiliser Production)

Optimum overall production volume = 540,294 t/yr

Corresponding (ε-constraint) effluent discharge rate = 41,172 t/yr

RM Production		Granulation		Distribution			
(A)	103,882	$(NPK)^{1}_{GR}$	414,920	$(NPK)^{1}_{S1}$	-	$(NPK)^{1}_{S5}$	83,823
$(A)_{AN}$	23,514	$(NPK)^{2}_{GR}$	39,431	$(NPK)^{2}_{S1}$	-	$(NPK)^{2}_{S5}$	17,198
$(A)_{NA}$	61,056	$(NPK)^{3}_{GR}$	1,444	$(NPK)^{3}_{S1}$	6,630	$(NPK)^{3}_{S5}$	12,459
$(A)_{NP}$	19,312	$(NPK)^{4}_{GR}$	1,047	$(NPK)^{4}_{S1}$	2,438	$(NPK)^{4}_{S5}$	1,309
(S)	41,038	$(NPK)^{5}_{GR}$	64,797	$(NPK)^{5}_{S1}$	-	$(NPK)^{5}_{S5}$	33,496
$(S)_{NP}$	39,026	$(NPK)^{6}_{GR}$	4,875	$(NPK)^{6}_{S1}$	-	$(NPK)^{6}_{S5}$	4,847
(W)	136,674	$(NPK)^{7}_{GR}$	3,122	$(NPK)^{7}_{S1}$	2,880	$(NPK)^{7}_{S5}$	34,763
(RP)	120,913	$(NPK)^{9}_{GR}$	33	$(NPK)^{9}_{S1}$	398	$(NPK)^{9}_{S5}$	317
$(RP)_{NP}$	57,307	$(NPK)^{10}_{GR}$	-	$(NPK)^{10}_{S1}$	-	$(NPK)^{10}_{S5}$	5,275
$(O)_{NA}$	262,656	$(NPK)^{11}_{GR}$	3,884	$(NPK)^{11}_{S1}$	-	$(NPK)^{11}_{S5}$	2,784
$(W)_{NA}$	60,288	$(NPK)^{13}_{GR}$	6,094	$(NPK)^{1}_{S2}$	850	$(NPK)^{1}_{S6}$	83,823
$(NA)_{NA}$	384,000	$(NPK)^{14}_{GR}$	647	$(NPK)^{2}_{S2}$	259	$(NPK)^{2}_{S6}$	-
$(NA)_{AN}$	75,067			$(NPK)^{3}_{S2}$	-	$(NPK)^{3}_{S6}$	12,459
$(W)_{AN}$	11,187			$(NPK)^{4}_{S2}$	-	$(NPK)^{4}_{S6}$	1,281
$(HF)_{NP}$	2,292			$(NPK)^{5}_{S2}$	-	$(NPK)^{5}_{S6}$	-
$(Ca)_{NP}$	103,438			$(NPK)^{6}_{S2}$	-	$(NPK)^{6}_{S6}$	-
$(W)_{NP}$	28,653			$(NPK)^{7}_{S2}$	-	$(NPK)^{7}_{S6}$	-
$(NA)_{NP}$	121,433			$(NPK)^{9}_{S2}$	-	$(NPK)^{9}_{S6}$	-
$(NA)_{CN}$	187,500			$(NPK)^{10}_{S2}$	-	$(NPK)^{10}_{S6}$	-
$(AN)_{AN}$	87,395			$(NPK)^{11}_{S2}$	-	$(NPK)^{11}_{S6}$	-
$(L)_{CN}$	11,220			$(NPK)^{1}_{S3}$	-	$(NPK)^{1}_{S7}$	-
$(NPH)_{NP}$	160,000			$(NPK)^{2}_{S3}$	-	$(NPK)^{2}_{S7}$	-
$(S)_{SP}$	2,013			$(NPK)^{3}_{S3}$	-	$(NPK)^{3}_{S7}$	-
$(SP)_{SP}$	154,818			$(NPK)^{4}_{S3}$	1,309	$(NPK)^{4}_{S7}$	-
$(RP)_{SP}$	63,607			$(NPK)^{5}_{S3}$	33,496	$(NPK)^{5}_{S7}$	-
$(W)_{SP}$	31,893			$(NPK)^{6}_{S3}$	4,847	$(NPK)^{6}_{S7}$	-
$(W)_{CN}$	15,840			$(NPK)^{7}_{S3}$	34,763	$(NPK)^{7}_{S7}$	-
$(CD)_{CN}$	38,880			$(NPK)^{9}_{S3}$	317	$(NPK)^{9}_{S7}$	-

$(CN)_{CN}$	144,000			$(NPK)^{10}_{S3}$	5,275	$(NPK)^{10}_{S7}$	-
$(NPH)_{GR}$	266,171			$(NPK)^{11}_{S3}$	2,784	$(NPK)^{11}_{S7}$	-
$(AN)_{GR}$	15,395			$(NPK)^{1}_{S4}$	67,856	$(NPK)^{1}_{S8}$	-
$(CN)_{ANCN}$	120,000			$(NPK)^{2}_{S4}$	13,922	$(NPK)^{2}_{S8}$	-
$(AN)_{ANCN}$	72,000			$(NPK)^{3}_{S4}$	5,425	$(NPK)^{3}_{S8}$	-
$(ANCN)_{ANCN}$	192,000			$(NPK)^{4}_{S4}$	1,060	$(NPK)^{4}_{S8}$	-
$(NPH)_{FBG}$	24,000			$(NPK)^{5}_{S4}$	27,108	$(NPK)^{5}_{S8}$	-
$(CN)_{FBG}$	24,000			$(NPK)^{6}_{S4}$	3,923	$(NPK)^{6}_{S8}$	-
$(SP)_{GR}$	154,818			$(NPK)^{7}_{S4}$	8,892	$(NPK)^{7}_{S8}$	-
(K_{GR})	103,910			$(NPK)^{9}_{S4}$	323	$(NPK)^{9}_{S8}$	-
				$(NPK)^{10}_{S4}$	5,345	$(NPK)^{10}_{S8}$	-
				$(NPK)^{11}_{S4}$	2,826	$(NPK)^{11}_{S8}$	-
				$(NPK)^{13}_{S4}$	12,674	$(NPK)^{13}_{S3}$	-
				$(NPK)^{14}_{S4}$	860	$(NPK)^{14}_{S8}$	-
			540,294		246,460		293,834

7.3.3 Conclusion

The fact that the optimum solution set yielded by the 'Compromise Optimum Solution' technique, Table 7-12, as put forward by Bit et al. (1992), is precisely the same as the optimum solution set yielded by the $\in$-constraint methodology, originally put forward by Haimes and Chankong, (1979) is valid confirmation that $\in$-constraint solution set is, indeed, optimal

7.4 Variations in Operational Planning and Uncertainty

Since the intention behind the proposed 'Planning and Optimisation of a Supply Chain Network under Uncertainty' methodology was to cater for any prevailing conditions of operational planning and uncertainty in a supply chain operating environment, it would be very useful to test the proposed methodology against any possible variations in this regard.

The instance of planning and uncertainty considered in the case study of this research report was represented by *multi-objectivity* and *fuzziness*, where *multi-objectivity* was represented by the need to *i*) maximise production and distribution of the entire range of NPK fertiliser in accordance with market demand, and *ii*) simultaneously minimise the generation and discharge of hazardous gaseous effluent, i.e. hydrogen fluoride $(HF)_{NP}$ discharge from the Nitrophosphate unit and carbon dioxide discharge $(CD)_{CN}$ from the Calcium Nitrate (CN) reactor. *Fuzziness* ($<$, $\leq$, $>$ or $\geq$) was reflected in the quantitative operational uncertainties surrounding the various production operational processes, e.g. granulation - $0.1(NPH)^{\mathrm{i}}_{GR} + 0.35(AN)^{i}_{GR} \leq [N]^{\mathrm{i}}(LA)^{I}$, as well as the various distribution processes, e.g. $(NPK)^{II}_{S5} \leq (1.98)(1406) \leq 2{,}784$. Since varied incidences of *fuzziness* are abundant in most operational projects, the decision was taken to consider only variations in *multi-objective* planning and/or *stochastic* uncertainty in this research, and then to assess whether the proposed methodology can withstand such variation. The manner in which this will be done will be to gradually introduce and/or remove variable planning and operational uncertainty from the base agri-fertiliser supply chain under uncertainty model, Table 7-6, in the manner put forward in Table 7-13. The above-mentioned methodology assessment will mainly take the form of checking whether a consequent set of optimum results falls within the realms of reality or not. Table 7-13 was constructed using a logical sequence of uncertainty events.

Table 7-13: Schedule of introduction/removal of operational planning and uncertainty into/from case study model

1	Case study model under multi-objective planning and fuzzy operational uncertainty
2	Introducing stochastic uncertainty by converting fuzzy demand uncertainty into discrete scenarios of probabilistic demand uncertainty
3	Introducing an additional element of multi-objective planning by means of an ε-constraint
4	Removing two of the, now, three instances of multi-objective planning as well as removing probabilistic uncertainty, and thereby leaving just a single objective requirement and fuzzy demand uncertainty.

The more detailed ramifications of effecting these changes in each of the supply chain planning and operational uncertainty scenarios, listed in Table 7-13, will be discussed below, if need be:

7.4.1 Introducing stochastic uncertainty by creating discrete demand scenarios of probabilistic uncertainty

The distribution of the various specific blends of *NPK* fertiliser for crop-type *i* at a particular site, S_n, i.e. $(NPK)^i{}_{Sn}$ is dependent on the crop-type land area, $(LA)^I$, the recommended fertilisation rate for that particular crop, $(NPK)^i$ and, of course, on the probability of demand for that particular fertiliser at that particular site, i.e. $(p)^i{}_{Sn}$, The relationship, according to para. 7.2.7.1.2, is, $(NPK)^i{}_{Sn} \leq (p)^i{}_{Sn}(LA)^i(NPK)^I$ t/ha.

For the purposes of the case study, it was assumed that $(p)^i{}_{Sn} = 1$ across all crop-types and across all sites, S_n. However, for the sake of this specific exercise, certain probabilistic uncertainty demand scenarios will be assumed for all crop-types at all sites, Table 7-14

Table 7-14: Assumed probabilistic demand scenarios - crops i and sites, Sn

	Crop demand probability, *p* (%) at site, S_n							
Crop-Type, *i*	S_1	S_2	S_3	S_4	S_5	S_6	S_7	S_8
Maize	80	95	60	75	115	110	90	85
Wheat	75	80	45	105	120	105	95	80
Sunflower	90	110	75	95	100\	90	110	97
Skybeans	85	90	55	60	80	75	95	95
Sugar Cane	100	110	80	70	105	95	115	110
Lucerne	110	115	95	105	120	115	100	95
Other Pastures	60	85	100	100	130	120	110	105
Tobacco	100	95	95	95	90	115	105	100
Cotton	80	75	90	95	85	90	110	95
Vegetables	110	95	95	100	95	100	105	100
Potatoes	85	85	95	110	100	105	100	95
Citrus	75	75	85	90	90	85	95	100
Subtropical	60	75	90	85	75	110	125	130
Fruits/Nuts	90	65	110	120	110	100	115	120

This new probability based demand distribution can now enable a recalculation of the new distribution constraints according to the NPK distribution relationship, $(NPK)^i{}_{Sn} \leq (p)^i{}_{Sn}(LA)^i(NPK)^I$ t/ha, which will be the same as multiplying the distribution limits in Table 7-6 by the respective probability factors in Table 7-14 and this is achieved in Table 7-15.

Table 7-15: Demand probability - amended distribution limits

S1	**S3**	**S5**
$(NPK)^{1}_{S1} < 155,828(0.80) < 124,662$	$(NPK)^{1}_{S3} < 83,823(0.60) < 50,293$	$(NPK)^{1}_{S5} < 83,823(1.15) < 96,396$
$(NPK)^{2}_{S1} < 31,970(0.75) < 23,978$	$(NPK)^{2}_{S3} < 17,198(0.45) < 7,739$	$(NPK)^{2}_{S5} < 17,198(1.20) < 20,638$
$(NPK)^{3}_{S1} < 12,459(0.90) < 11,213$	$(NPK)^{3}_{S3} < 12,459(0.75) < 9,344$	$(NPK)^{3}_{S5} < 12,459(1.00) < 12,549$
$(NPK)^{4}_{S1} < 2,438(0.85) < 2,072$	$(NPK)^{4}_{S3} < 1,309(0.55) < 720$	$(NPK)^{4}_{S5} < 1,309(0.80) < 1,047$
$(NPK)^{5}_{S1} < 62,260(1.00) < 62,260$	$(NPK)^{5}_{S3} < 33,496(0.80) < 26,797$	$(NPK)^{5}_{S5} < 33,496(1.05) < 35,170$
$(NPK)^{6}_{S1} < 9,012(1.10) < 9,913$	$(NPK)^{6}_{S3} < 4,847(0.95) < 4,605$	$(NPK)^{6}_{S5} < 4,847(1.20) < 5,816$
$(NPK)^{7}_{S1} < 31,975(0.60) < 19,185$	$(NPK)^{7}_{S3} < 34,763(1.00) < 34,763$	$(NPK)^{7}_{S5} < 34,763(1.30) < 45,192$
$(NPK)^{8}_{S1} < 1,094(1.00) < 1,094$	$(NPK)^{8}_{S3} < 586(0.95) < 557$	$(NPK)^{8}_{S5} < 586(0.90) < 527$
$(NPK)^{9}_{S1} < 31,977(0.80) < 25,582$	$(NPK)^{9}_{S3} < 317(0.90) < 285$	$(NPK)^{9}_{S5} < 317(0.85) < 269$
$(NPK)^{10}_{S1} < 9,779(1.10) < 10,756$	$(NPK)^{10}_{S3} < 5,275(0.95) < 5,011$	$(NPK)^{10}_{S5} < 5,275(0.95) < 5,011$
$(NPK)^{11}_{S1} < 5,160(0.85) < 4,386$	$(NPK)^{11}_{S3} < 2,784(0.95) < 2,645$	$(NPK)^{11}_{S5} < 2,784(1.00) < 2,784$
S2	**S4**	**S6**
$(NPK)^{1}_{S2} < 83,823(0.95) < 79,632$	$(NPK)^{1}_{S4} < 67,8563(0.75) < 5,089$	$(NPK)^{1}_{S6} < 83,823(1.10) < 92,205$
$(NPK)^{2}_{S2} < 17,198(0.80) < 13,758$	$(NPK)^{2}_{S4} < 13,922(1.05) < 14,618$	$(NPK)^{2}_{S6} < 17,198(1.05) < 18,058$
$(NPK)^{3}_{S2} < 12,459(1.10) < 13,705$	$(NPK)^{3}_{S4} < 5,425(0.95) < 5,154$	$(NPK)^{3}_{S6} < 12,459(0.90) < 11,213$
$(NPK)^{4}_{S2} < 1,309(0.90) < 1,178$	$(NPK)^{4}_{S4} < 1,060(0.60) < 636$	$(NPK)^{4}_{S6} < 1,309(0.75) < 982$
$(NPK)^{5}_{S2} < 33,496(0.80) < 26,797$	$(NPK)^{5}_{S4} < 27,108(0.70) < 18,976$	$(NPK)^{5}_{S} < 33,496(0.95) < 31,821$
$(NPK)^{6}_{S2} < 4,847(1.15) <$	$(NPK)^{6}_{S4} < 3,923(1.05) <$	$(NPK)^{6}_{S6} < 4,847(1.15) <$

5,574	412	5,574
$(NPK)^{7}_{S2} < 34{,}763(0.85) <$ 29,549	$(NPK)^{7}_{S4} < 8{,}892(1.00) <$ 8,892	$(NPK)^{7}_{S6} < 34{,}763(1.20) <$ 41,716
$(NPK)^{8}_{S2} < 586(0.95) < 557$	$(NPK)^{8}_{S4} < 595(0.95) < 565$	$(NPK)^{8}_{S6} < 586(1.15) < 674$
$(NPK)^{9}_{S2} < 317(0.75) < 238$	$(NPK)^{9}_{S4} < 323(0.95) < 307$	$(NPK)^{9}_{S6} < 317(0.90) < 285$
$(NPK)^{10}_{S2} < 5{,}275(0.95) <$ 5,011	$(NPK)^{10}_{S4} < 5{,}345(1.00) <$ 5,345	$(NPK)^{10}_{S6} < 5{,}275(1.00) <$ 5,275
$(NPK)^{11}_{S2} < 2{,}784(0.85) <$ 2,366	$(NPK)^{11}_{S4} < 2{,}826(1.10) <$ 3,109	$(NPK)^{11}_{S6} < 2{,}784(1.05) <$ 2,923
	$(NPK)^{12}_{S4} < 4{,}376(0.90) <$ 3,938	
	$(NPK)^{13}_{S4} < 12{,}674(0.85) <$ 10,773	
	$(NPK)^{14}_{S4} < 860(1.20) <$ 1,032	
	$(NPK)^{15}_{S4} < 20{,}900(0.85) <$ 17,765	

The effects of these changes are discussed in section 7.4.5.

7.4.2 Maximising the production of ammonium nitrate $(AN)_{AN}$ whilst still retaining the probabilistic distribution of NPK fertiliser

An additional element of multi-objective planning will be achieved through the introduction of an additional $\in$-constraint in the agri-fertiliser production and distribution MILP. It was decided that such $\in$-constraint would take the form of maximising the production of ammonium nitrate, i.e. $(AN)_{AN}$, despite the fact that it is already a key raw material in the production of NPK fertiliser. Ammonium nitrate can either be used as an explosive or as a fertiliser, in its own right.

Such additional $\in$-constraint would take the form of $(AN)_{AN} \leq \text{Max}[(AN)_{AN}]$, where $\text{Max}[(AN)_{AN}]$, as a single objective = 224,000 t/yr. Therefore the new $\in$-constraint parameters in the Microsoft Excel Solver ver. 2010, i.e. Premium Solver program, where the primary objective function is Max $[\Sigma(NPK)_{DIS}]$ and the two $\in$-constraints are:

i) $[(HF)_{NP} + (CD)_{CN}] \geq \text{Min}[(HF)_{NP} + (CD)_{CN}] \geq 38{,}880$ t/yr

ii) $(AN)_{AN} \leq \text{Max}[(AN)_{AN}] \leq 224{,}000$ t/yr

Again, the effects of these changes are given and discussed in section 7.4.5.

7.4.3 Minimising the generation and discharge of hazardous effluent whilst retaining probabilistic demand distribution of NPK fertiliser

It will be an interesting exercise to minimise the generation and discharge of hazardous gaseous effluent, i.e. hydrogen fluoride $(HF)_{NP}$ from the Nitrophosphate unit and carbon dioxide $(CD)_{CN}$ from the CN (Calcium Nitrate) reactor, to see what impact that has on the primary production objective, i.e. the generation of multiple blends of NPK fertiliser in accordance with market demand. In this particular case, the objective function of the MILP will be $\text{Min}[(HF)_{NP} + (CD)_{CN}]$.

Again, the effects of these changes are given and discussed in section 7.4.5.

7.4.4 Maximising the production of Ammonium Nitrate whilst replacing the probabilistic demand of NPK fertiliser with the previous fuzzy demand of NPK fertiliser

It will also be interesting to see whether the replacement of probabilistic demand for (*NPK*) fertiliser, with the original case study requirement of fuzzy demand for such fertiliser, will have much impact on the optimum operating results.

Again, the effects of these changes are given and discussed in section 7.4.5.

7.4.5 Discussion of Variation Results

It was decided to display the results of the effects of the changing nature of all these supply chain operational planning and/or operational uncertainty variations in sequential tabular format, Table 7-16, so that the nature of these changing effects can be properly viewed and discussed.

Table 7-16: The effects of changing operational planning and uncertainty on supply chain performance

Categories ε-constraints	Case Study - Max(*NPK*)	Max(*NPK*) + Demand Probability	Max(*AN*) + Demand Probability	Min(Eff) + Demand Probability	Max(AN)
$(NPK)_{DIS}$	540,294	540,294	227,232	55,776	227,232
$(AN)_{AN}$			224,000	72,000	224,000
$(HF)_{NP}$ + $(CD)_{CN}$	41,172	41,172	38,880	38,880	38,880
Variables					
(A)	103,882	103,882	120,673	59,018	120,673
$(A)_{AN}$	23,514	23,514	60,268	19,372	60,268
$(A)_{NA}$	61,056	61,056	60,405	39,646	60,405
$(A)_{NP}$	19,312	19,312	-	-	-
(S)	41,038	41,038	-	-	-
$(S)_{NP}$	39,026	39,026	-	-	-
(W)	136,674	136,674	75,485	54,987	75,485
(RP)	120,913	120,913	-	-	-
$(RP)_{NP}$	57,307	57,307	-	-	-
$(O)_{NA}$	262,656	262,656	259,854	170,551	259,854
$(W)_{NA}$	60,288	60,288	59,645	39,147	59,645
$(NA)_{NA}$	384,000	384,000	379,904	249,344	379,904
$(NA)_{AN}$	75,067	75,067	192,404	61,844	192,404
$(W)_{AN}$	11,187	11,187	28,672	9,216	28,672
$(HF)_{NP}$	2,292	2,292	-	-	-
$(Ca)_{NP}$	103,438	103,438	-	-	-
$(W)_{NP}$	28,653	28,653	-	-	-
$(NA)_{NP}$	121,433	121,433	-	-	-
$(NA)_{CN}$	187,500	187,500	187,500	187,500	187,500
$(AN)_{AN}$	87,395	87,395	224,000	72,000	224,000
$(L)_{CN}$	11,220	11,220	11,220	11,220	11,220
$(NPH)_{NP}$	160,000	160,000	-	-	-
$(S)_{SP}$	2,013	2,013	-	-	-
$(SP)_{SP}$	154,818	154,818	-	-	-
$(RP)_{SP}$	63,607	63,607	-	-	-
$(W)_{SP}$	31,893	31,893	-	-	-
$(W)_{CN}$	15,840	15,840	15,840	15,840	15,840
$(CD)_{CN}$	38,880	38,880	38,880	38,880	38,880
$(CN)_{CN}$	144,000	144,000	144,000	144,000	144,000
$(NPH)_{GR}$	266,171	266,171	75,232	55,776	75,232
$(AN)_{GR}$	15,395	15,395	152,000	-	152,000
$(CN)_{ANCN}$	120,000	120,000	120,000	120,000	120,000
$(AN)_{ANCN}$	72,000	72,000	72,000	72,000	72,000
$(ANCN)_{ANCN}$	192,000	192,000	192,000	192,000	192,000
$(NPH)_{FBG}$	24,000	24,000	24,000	24,000	24,000
$(CN)_{FBG}$	24,000	24,000	24,000	24,000	24,000
$(SP)_{GR}$	154,818	154,818	-	-	-
(K_{GR})	103,910	103,910	-	-	-
$(NPK)^{1}_{S1}$	-	-	-	-	-

$(NPK)^{2}_{S1}$	-	-	-	-	-
$(NPK)^{3}_{S1}$	-	-	-	-	-
$(NPK)^{4}_{S1}$	-	-	-	-	-
$(NPK)^{5}_{S1}$	-	-	-	-	-
$(NPK)^{6}_{S1}$	-	-	-	-	-
$(NPK)^{7}_{S1}$	-	-	-	-	-
$(NPK)^{9}_{S1}$	-	-	-	-	-
$(NPK)^{10}_{S1}$	-	-	-	-	-
$(NPK)^{11}_{S1}$	5,160	4,386	-	-	-
$(NPK)^{1}_{S2}$	-	-	-	-	-
$(NPK)^{2}_{S2}$	-	-	-	-	-
$(NPK)^{3}_{S2}$	-	-	-	-	-
$(NPK)^{4}_{S2}$	1,309	1,178	-	-	-
$(NPK)^{5}_{S2}$	33,496	26,297	-	-	-
$(NPK)^{6}_{S2}$	4,847	5,574	-	-	-
$(NPK)^{7}_{S2}$	34,763	29,549	-	-	-
$(NPK)^{9}_{S2}$	317	238	-	-	-
$(NPK)^{10}_{S2}$	5,275	5,011	-	-	-
$(NPK)^{11}_{S2}$	-	-	-	-	-
$(NPK)^{1}_{S3}$	83,823	50,293	-	-	-
$(NPK)^{2}_{S3}$	17,198	7,739	-	-	-
$(NPK)^{3}_{S3}$	12,459	9,344	-	-	-
$(NPK)^{4}_{S3}$	1,309	1,309	-	-	-
$(NPK)^{5}_{S3}$	33,496	26,797	-	-	-
$(NPK)^{6}_{S3}$	4,847	4,605	-	-	-
$(NPK)^{7}_{S3}$	34,763	34,763	-	-	-
$(NPK)^{9}_{S3}$	317	285	-	-	-
$(NPK)^{10}_{S3}$	5,275	5,275	-	-	-
$(NPK)^{11}_{S3}$	2,784	2,645	-	-	-
$(NPK)^{1}_{S4}$	67,856	50,892	-	-	-
$(NPK)^{2}_{S4}$	13,922	14,618	-	-	-
$(NPK)^{3}_{S4}$	5,425	5,154	-	-	-
$(NPK)^{4}_{S4}$	1,060	636	-	-	-
$(NPK)^{5}_{S4}$	27,108	18,976	-	-	-
$(NPK)^{6}_{S4}$	3,923	412	-	-	-
$(NPK)^{7}_{S4}$	8,892	8,892	-	-	-
$(NPK)^{9}_{S4}$	323	307	-	-	-
$(NPK)^{10}_{S4}$	5,345	5,345	-	-	-
$(NPK)^{11}_{S4}$	2,826	3,109	-	-	-
$(NPK)^{13}_{S4}$	12,674	10,773	-	-	-
$(NPK)^{14}_{S4}$	860	1,032	-	-	-
$(NPK)^{1}_{S5}$	83,823	96,396	-	-	-
$(NPK)^{2}_{S5}$	17,198	20,638	-	-	-
$(NPK)^{3}_{S5}$	7,621	12,459	-	-	-
$(NPK)^{4}_{S5}$	-	1,047	-	-	-
$(NPK)^{5}_{S5}$	-	35,170	-	-	-
$(NPK)^{6}_{S5}$	-	5,816	-	-	-
$(NPK)^{7}_{S5}$	-	33,334	-	-	-
$(NPK)^{9}_{S5}$	-	-	-	-	-

$(NPK)^{10}{}_{S5}$	-	-	-	-	-
$(NPK)^{11}{}_{S5}$	-	-	-	-	-
$(NPK)^{1}{}_{S6}$	-	-	-	-	-
$(NPK)^{2}{}_{S6}$	-	-	-	-	-
$(NPK)^{3}{}_{S6}$	-	-	-	-	-
$(NPK)^{4}{}_{S6}$	-	-	-	-	-
$(NPK)^{5}{}_{S6}$	-	-	27,581	-	27,581
$(NPK)^{6}{}_{S6}$	-	-	-	-	-
$(NPK)^{7}{}_{S6}$	-	-	-	-	-
$(NPK)^{9}{}_{S6}$	-	-	-	-	-
$(NPK)^{10}{}_{S6}$	-	-	-	-	-
$(NPK)^{11}{}_{S6}$	-	-	-	-	-
$(NPK)^{1}{}_{S7}$	-	-	-	55,776	-
$(NPK)^{2}{}_{S7}$	-	-	17,198	-	17,198
$(NPK)^{3}{}_{S7}$	-	-	12,459	-	12,459
$(NPK)^{4}{}_{S7}$	-	-	1,309	-	1,309
$(NPK)^{5}{}_{S7}$	-	-	33,496	-	33,496
$(NPK)^{6}{}_{S7}$	-	-	4,847	-	4,847
$(NPK)^{7}{}_{S7}$	-	-	34,763	-	34,763
$(NPK)^{9}{}_{S7}$	-	-	317	-	317
$(NPK)^{10}{}_{S7}$	-	-	5,275	-	5,275
$(NPK)^{11}{}_{S7}$	-	-	2,784	-	2,784
$(NPK)^{1}{}_{S8}$	-	-	67,856	-	67,856
$(NPK)^{2}{}_{S8}$	-	-	13,922	-	13,922
$(NPK)^{3}{}_{S8}$	-	-	5,425	-	5,425
$(NPK)^{4}{}_{S8}$	-	-	-	-	-
$(NPK)^{7}{}_{S8}$	-	-	-	-	-
$(NPK)^{9}{}_{S8}$	-	-	-	-	-
$(NPK)^{10}{}_{S8}$	-	-	-	-	-
$(NPK)^{11}{}_{S8}$	-	-	-	-	-
$(NPK)^{13}{}_{S3}$	-	-	-	-	-
$(NPK)^{14}{}_{S8}$	-	-	-	-	-
Total $(NPK)^{i}{}_{Sn}$	540,294	540,294	227,232	55,776	227,232
$(FBG)_{S1}$	-	-	48,000	-	48,000
$(FBG)_{S2}$	-	-	-	-	-
$(FBG)_{S3}$	-	-	-	-	-
$(FBG)_{S4}$	-	-	-	48,000	-
$(FBG)_{S5}$	-	-	-	-	-
$(FBG)_{S6}$	-	-	-	-	-
$(FBG)_{S7}$	-	-	-	-	-
$(FBG)_{S8}$	48,000	48,000	-	-	-

It is clear from a perusal of Table 7-16 that there is remarkable consistency between the various sets of optimum results presented, which correspond to the various combinations of operational planning and uncertainty discussed.

For example, considering the first two cases, i.e. the case of fuzzy demand of NPK fertiliser, $(NPK)_{Dis}$, and the case of probabilistic demand for $(NPK)_{Dis}$,, they both yield identical production profiles and differ only, as to be expected, in their distribution profiles. Similarly, in the case where the production of ammonium nitrate, $(AN)_{AN}$, is maximised under fuzzy market demand conditions, identical profile results are obtained in the case where the probabilistic demand of $(NPK)_{Dis}$ fertiliser is considered. Lastly, there is also consistency between the two cases of i) maximising the production of $(AN)_{AN}$ and ii) minimising the generation and discharge of hazardous effluent, $(HF)_{NP}$ + $(CD)_{CN}$, in that, since the production of $(AN)_{AN}$ has absolutely nothing to do with the generation of effluent, the corresponding effluent figures are exactly the same at 38,880 t/yr.

Chapter 8

STRENGTHS AND WEAKNESSES

The original rationale for the development of this new 'Planning and Optimisation of a Supply Chain under Uncertainty' methodology was because it became increasingly apparent, especially in the late 20th, and early 21st, century that more and more international research work was being done into the various aspects of 'Supply Chain, under uncertainty, Planning/Optimisation', i.e. the nature of supply chains, the nature of supply chain uncertainty and the meaning of supply chain planning/optimisation However it became clear that, since such previous international research works were neither coordinated nor conforming to a central or common theme, there was no binding rationale between these various aspects of 'Supply Chain, under uncertainty, Planning/Optimisation' research. Therefore it was decided to create such binding rationale by developing an overall supply chain planning/optimisation' under uncertainty methodology that could cater for any supply chain infrastructure and also cater for any (combinational) instances of uncertainty.

This was achieved and it now becomes necessary to discuss the strengths and weaknesses of the derived methodology against these constraints.

8.1 Strengths

1) Puts into logical and sequential perspective a lot of the previous work into supply, chain under uncertainty, planning/optimisation research.

2) The derived methodology can cater for any combined prevailing instances of operational planning and/or uncertainty.

3) For the first time ever, any and all aspects of supply chain planning and operation are catered for in one methodology.

8.2 Weaknesses

1) There is only one real key weakness in the derived methodology and that is that it caters only for the operational aspects of supply chain performance but not for any of the design aspects of supply chain operational performance. A key example of this is the design of the supply chain infrastructure, which includes the size and capacity of the production/manufacturing facilities, the number and capacities of warehouses, distribution centres and retail sales outlets and lastly, the orientation of the entire supply chain infrastructure, in terms of location and routes, towards market demand.

Chapter 9

REMARKS AND CONCLUSION

This study has effectively demonstrated the derivation and an application of a generic 'Planning and Optimisation Methodology for a Supply Chain, under Uncertainty', in a supply chain environment., The derived methodology was also validated by successfully comparing optimum solution results with those generated by a prior, uncertainty-specific supply chain optimisation technique,. Additionally, it was also successfully demonstrated that the derived methodology could handle variations in uncertainty.

The development of the methodology itself was very much based on the research and analysis of previously researched supply chain, under uncertainty, optimisation techniques. Such existing techniques were either single component in nature, occasionally binary and very occasionally ternary based, but never did they consider only those elements of operational uncertainty that were relevant to a particular supply chain environment. They also did not provide for mechanisms for handling binary logic in the determination of optimum configurations for supply chain networks vis-à-vis market demand. Further, application of the ε-constraint method, in the case of multiple objectives, had only been occasionally utilised to generate Pareto optimal supply chain solutions Therefore, the intention of this thesis was to cater for the possible planning of supply chain networks that were also subject to any prevailing conditions of operational uncertainty, be they single-case or binary in nature, and by utilising the most effective and proven technology available.. Consequently, this would amount to the determination of a *'realistically planned'* optimum supply chain operational solution for such environments since all prevailing conditions of operational uncertainty would have been accommodated, and any multi-objective planning requirements would also have been accommodated.

The supply chain case study was very interesting in itself because not only did it exhibit one (fuzzy) prevailing, out of a possible two (fuzzy, stochastic), conditions of operational uncertainty but the operation was also characterised by the need of two performance objectives. Another point of interest was the specified relationships between the crop-type and the associated NPK requirement in terms of the stipulated individual N, P, K, concentrations. Such relationships made it possible to optimise both the production and raw material stream requirements. The effectiveness of this methodology was demonstrated by the nature of results achieved, i.e. 100% achievement of optimal NPK fertiliser production and distribution capability and further, a 94.4% achievement of optimum effluent discharge conditions.

Chapter 10

FUTURE POSSIBLE RESEARCH AND FURTHER APPLICATION

This is a reflective chapter that considers any further possible areas of supply chain, under uncertainty, research and also looks at any further possible areas of application. The former point, research, is considered initially because it is primarily based on the identified strengths and weaknesses of the derived methodology, chapter 9, whereas the latter point, application, is primarily based on previously researched (References, 14) applications of the underlying MILP/MINLP (Mixed integer Linear/Non-Linear Programming), under uncertainty, technology.

10.1 Future possible research

It would be useful if the derived supply chain, under uncertainty, operational optimisation methodology could be expanded to include the operational implications of any supply chain infrastructure design change requirements. Such supply chain infrastructure design change requirements could typically include either quantitative and/or qualitative enhancements to the supply chain production or manufacturing facilities and/or the number and capacities of warehouses, distribution centres and retail outlets and the orientation of the entire supply chain infrastructure towards market demand. This latter point could typically be qualified in terms of permissible routes and locations.

10.1.1 Quantitative enhancements to methodology

The derived 'Supply Chain, under Uncertainty, Optimisation' methodology, as defined and described in chapter 6, *'Specification of a Planning and Optimisation Methodology for a Supply Chain Network under Uncertainty'* may be enhanced to accommodate the desired quantitative and/or qualitative design changes to the supply chain network under consideration.

Any **Quantitative** enhancements to a supply chain network design would typically involve any desired changes to the number and/or capacities of the various supply chain components, i.e. number and capacities of production/manufacturing facilities and distribution centres as well as any retail sales outlets, involved in a supply chain network.

Once the illustrated SCN diagram is correctly updated to reflect the required quantitative changes, the MILP/MINLP program(s) would also have to be modified to accurately reflect the required operational changes. Such program(s) would then be solved to obtain the required optimum solution.

10.1.2 Qualitative improvements to methodology

Any **Qualitative** enhancements to a supply chain network might embrace any new or modified supply chain components to improve product quality and/or reduce product rejection rates. Additionally or alternatively, a qualitative enhancement could include the specification or rejection of certain supply chain routes for specific products. Again, the MILP/MINLP program(s) would also have to be modified to accurately accommodate the required supply chain architectural changes. Such program(s) would then be solved to obtain the required optimum solution

10.2 Further possible application

Much supply chain architectural and operational research has been conducted over the last few years, and especially since the beginning of the 21st century. The operational basis of these various research initiatives has varied from mostly logical component variations of the traditional supply chain infrastructure, i.e. Procurement, Inventory, Production/Manufacturing, Warehousing, Distribution, and Retail Sales to incorporate only, say, Production, Warehousing, Distribution and Sales or to incorporate only, say, Inventory, Production, Warehousing and Distribution. However, there have been other supply chain variations that have dealt with the flow of information rather than that of physical goods.

One example of these information flow type supply chains could be customer service centres where different information flow types could typically include lodging an enquiry, lodging a complaint or registering a new account. Another example could be an Internet-based booking or reservation service where the various information flow types would typically include enquiries, reservations and complaints. As with physical goods supply chains, an informational supply chain network flow diagram would be constructed indicating all activity or decision points as well as all alternative information flow routes. Once all this is in place, it would then be subject to the normal MILP/MINLP programming and solution procedure.

Chapter 11

NOMENCLATURE

Table 11-1: Nomenclature

Category	**Description**
Indices/Sets	
Raw Materials	Ammonia (*A*), Sulphuric Acid (*S*), Rock phosphate (*RP*), Oxygen (*O*), Water (*W*), KCl (*K*), Limestone (*L*)
Intermediate Products	, e.g. Nitric Acid (*NA*), Nitrophos (*NP*), Superphosphates (*SP*), Ammonium Nitrate (*AN*), Calcium Nitrate (*CN*)
Final fertiliser products,	$(FBG)_i$, $(GR2)_j$, $(GR3)_k$ where *i, j, k* refer to the many various blends of NPK fertiliser exiting from the Granulators
S_n	Sites ($n = 1,\ldots, 8$) e.g. S_1, S_2
Variable Parameters	**Description**
(A)	Flowrate of NH_3 to the Fertiliser Production complex (t/yr)
$(A)_{AN}$	Flowrate of NH_3 to Ammonium Nitrate (*AN*) plant (t/yr)
$(A)_{NA}$	Flowrate of NH_3 to Nitric Acid (*NA*) plant (t/yr)
$(A)_{NP}$	Flowrate of NH_3 to Nitrophos (*NP*) plant (t/yr)
(S)	Flowrate of 98%Sulphuric acid to fertiliser complex (t/yr)
$(S)_{NP}$	Flowrate of 98%Sulphuric acid to *NP* plant (t/yr)
$(RP)_{NP}$	Flowrate of Phosphate rock to *NP* plant (t/yr)
$(A)_{NA}$	Flowrate of Compressed air *to NA* plant (t/yr)
$(O)_{NA}$	Flowrate of Oxygen *to NA* plant (t/yr)
$(W)_{NA}$	Flowrate of Water to *NA* plant (t/yr)
$(NA)_{NA}$	Flowrate of 59% Nitric Acid from *NA* Plant (t/yr)
$(NA)_{AN}$	Flowrate of 59% Nitric Acid from *NA* plant to *AN* plant (t/yr)
$(NA)_{NP}$	Flowrate of 59% Nitric Acid from *NA* plant to *NP* plant (t/yr)
$(AN)_{AN}$	Flowrate of 89% Ammonium Nitrate from *AN* plant (t/yr)
$(NA)_{CN}$	Flow of 59% Nitric Acid from NA plant to Calcium Nitrate (*CN*) plant (t/yr)

$(L)_{CN}$	Flowrate of Limestone ($CaCO_3$) to *CN* plant (t/yr)
$(W)_{CN}$	Release of water from *CN* plant (t/yr)
$(CD)_{CN}$	Release of Carbon Dioxide from *CN* plant (t/yr)
$(NPH)_{NP}$	Flowrate of Nitrophosphate from *NP* plant (t/yr)
$(W)_{NP}$	Flowrate of Water to *NP* plant (t/yr)
$(HF)_{NP}$	Hydrogen Fluoride discharge from *NP* plant (t/yr)
$(Ca)_{NP}$	$Ca(NO_3)_2.4H_2O$ discharge from *NP* plant (t/yr)
$(NA)_{NP}$	Incoming 59%Nitric Acid to the *NP* plant (t/yr)
$(S)_{SP}$	Flowrate of Sulphuric Acid to Superphosphates plant (t/yr)
$(SP)_{SP}$	Flow of Superphosphate from Superphosphates plant (t/yr)
$(RP)_{SP}$	Flowrate of Phosphate Rock to Superphosphates plant (t/yr)
(RP)	Total Flowrate of Phosphate Rock (t/yr)
W_{SP}	Flowrate of Water to Superphosphates plant (t/yr)
$(CN)_{CN}$	Flowrate of 58%Calcium Nitrate from *CN* plant (t/yr)
$(CN)_{ANCN}$	Flowrate of Calcium Nitrate to *ANCN* (explosives) plant (t/yr)
$(AN)_{ANCN}$	Flowrate of 89% *AN* to *ANCN* (explosives) plant (t/yr)
$(ANCN)_{ANCN}$	Flowrate of *ANCN* from *ANCN* Plant (explosives) plant (t/yr)
$(NP)_{FBG}$	Flowrate of Nitrophos to Fluidised Bed Granulator (t/yr)
$(CN)_{FBG}$	Flowrate of 58% Calcium Nitrate to Fluidised Bed Granulator (t/yr)
$(NP)^i_{GR}$	Flowrate of Nitrophos for *NPK* blend *i* to Granulator (t/yr)
$(AN)^i_{GR}$	Flowrate of 89% *AN* for *NPK* blend *i* to Granulator (t/yr)
$(SP)^i_{GR}$	Flowrate of Superphosphate for NPK blend *i* to Granulator (t/yr)
K^i_{GR}	Flow of *KCl* for *NPK* blend *i* to Granulator (t/yr)
$(NPK)^i_{GR}$	Flowrate of *NPK* Fertiliser of blend *i* from Granulator (t/yr)
$(NPK)^i_{Sn}$	Flowrate of *NPK* fertiliser to Site *n* (t/yr) of NPK blend *i*
$(LA)^i$	Land area involved for crop *i*, (ha)
$(p)^i_{Sn}$	Probability of complete demand for crop *i* at site, *n*, on account of environmental and/or economic conditions
$(FBG)_{Sn}$	Flowrate of fluidised bed granulated *CN* to Site *n* (t/yr)
$(FBG)_{DIS}$	Total flowrate of granulated *CN+NPH* from FBG to be distributed (t/yr)

Operational Constraints	**Description**
(LA)	Total Fertile Land Area involved = 4,241,000 ha
$(d)_{yr}$	Effective operational days pr annum (88%) = 320 (88% uptime)
$(NA)_{NA}$	Production flowrate of 59% Nitric Acid = 1200t/d
$(AN)_{AN}$	Flowrate of 89% Ammonium Nitrate from AN plant = 700t/d
$(ANCN)_{ANCN}$	Production capacity ≤ 280,000 t/yr
$(FBG)_{FBG}$	Production capacity of granulated CN from FBG = 150 t/day or ≤ 50,000 t/yr
$(SP)_{SP}$	Production capacity of Superphosphates plant = 450 t/day for SSP and 150 t/day for MCP ~ 500 t/day
$(CN)_{CN}$	500 t/day
$[P]^i$	Recommended application rate (kg/ha) of *P (phosphorous)* for crop type *i* (Table 7-4)
$[K]^i$	Recommended application rate (kg/ha) of *K (potassium)* for crop type *i* (Table 7-4)
$[N]^i$	Recommended application rate (kg/ha) of *N (nitrogen)* for crop type *I* (Table 7-4)
[NPK]	Equivalent daily production capacity of one Granul Granulator's 1 & 2 = 1,700 t/day

Chapter 12

APPENDICES

12.1 Transforming Probabilistic relationships into their Deterministic equivalents

It became evident, during preliminary research, that certain specialised programming techniques became necessary, especially in the formulation of certain stochastic-based MINLP's (Mixed Integer Non-Linear Programs), where certain constraints may be probabilistic in nature, e.g. a Normal or a Triangular Probability Distribution relationship. In cases such as these, in order to formulate a MINLP program, it becomes necessary to first transform such probabilistic relationships into their deterministic equivalents. Two such research papers were discovered in this regard, and these are:

i) 'A gradient algorithm for chance constrained nonlinear goal programming' – Lee and Olson, (1985) and,

ii) 'Non-normal Deterministic Equivalents and a Transformation in Stochastic Mathematical Programming' – Goicoechea and Duckstein (1987)

12.1.1.1 Chance constrained nonlinear goal programming through the use of gradient algorithms and incorporating probabilistic to deterministic transformation - Lee and Olson, (1985)

The fact that certain decision environments are frequently characterised by many conflicting objectives and apart from being described by many different levels of operational uncertainty, sometimes sampling information is available as a means of describing this uncertainty. Such description can be appropriately achieved in the form of chance constraints and, in conjunction with this, nonlinear programming provides the means of resolving the multiple, conflicting objectives. It is therefore proposed that a nonlinear programming algorithm, based upon the gradient method and optimal step length, is utilised for chance constrained goal programming models. The resulting algorithm was evaluated for general applicability, reliability, precision sensitivity to parameters, data preparational and computational effort and convergence and was subsequently found to require minimal preparational effort. It also displayed favourable computational time and achieved rapid convergence to the optimal solution, with exception in the case of models characterised by high degrees of' nonlinearity.

12.1.1.2 Non-normal Deterministic Equivalents and a Transformation in Stochastic Mathematical Programming – Goicoechea and Duckstein (1987)

In a paper by Goicoechea and Duckstein (1987), random variables of the continuous type are presented in two different ways, in a stochastic programming problem, to describe the treatment of objectives and constraints. Firstly, in the case of objective functions, an approach was presented for transforming them into their deterministic equivalents, and secondly, in the case of constraints, a general transformation approach was developed so that deterministic equivalents of constraints could be achieved within certain probability limits, and such an approach could be applied to continuous random variables as well. In the case of objective functions, the deterministic transformation of the stochastic program will yield a closed-form solution without resort to a Monte Carlo computer simulation. These approaches could also be extended to include stochastic problems with both discrete random variables as well as integer decision variables. A numeric example is also included.

12.2 NPK Ratio

[(1)]NPK: *Unlike the N number, the numbers for P and K do not reflect the amount of elemental phosphorus and potassium in the fertilizer. Rather they represent the amount of oxide in the form of* P_2O_5 *and* K_2O *that would be present in the fertilizer if all the elemental phosphorus and potassium were oxidized into these forms. The factors for converting from* P_2O_5 *and* K_2O *values to their respective P and K elemental values are as follows:*

- P_2O_5 consists of 56.4% oxygen and 43.6% elemental phosphorus. The percentage (mass fraction) of elemental phosphorus is 43.6% so P = 0.436 x P_2O_5
- K_2O consists of 17% oxygen and 83% elemental potassium. The percentage (mass fraction) of elemental potassium is 83% so K = 0.83 x K_2O
- Nitrogen values represent actual nitrogen content so these numbers do not need to be converted.

FSSA, FAO-UN, (2005)

REFERENCES

Aissi, H., Bazgan, C. & Vanderpooten, D. (2009). Min-max and min-max regret versions of combinatorial optimisation problems: A survey. *European Journal of Operational Research, 197*(2), 427-438.

Aliev, R.A., Fazlollahi, B., Guirimov, B.G. & Aliev, R.R (2007), Fuzzy-genetic approach to aggregate production-distribution planning in supply chain management, *Information Sciences,* 177(2007) 4241-4255

Al-Othman, W.B.E., Lababidi, H.M.S., Alatiqi, I.M., & Al-Shayji, K. (2008). Supply chain optimisation of petroleum organization under uncertainty in market demands and prices. *European Journal of Operational Research, 189*(3), 822-840.

Avalle, M., Chiandussi, G., (2007), Optimisation of a vehicle energy absorbing steel component with experimental validation, International Journal of Impact Engineering, Volume 34, Issue, Pages 843-858, ISSN 0734-743.

Baky, I.A. (2010), Solving multi-level multi-objective linear programming problems through fuzzy goal planning approach, *Applied Mathematical Modelling,* (34) 2377 - 2387

Baoding, L. (1996), Dependent-chance goal programming and its genetic algorithm based approach, *Mathematical and Computer Modelling,* Volume 24, Issue 7, Pages 43-52, ISSN 0895-7177.

Baoding L & Iwamura, K. (1998), Chance constrained programming with fuzzy parameters, *Fuzzy Sets and Systems Volume 94,* Issue 2, Pages 227-237, ISSN 0165-0114,

Baoding L & Iwamura K (1997), Modelling stochastic decision systems using dependent-chance programming, *European Journal of Operational Research,* Volume 101, Issue 1, Pages 193-203, ISSN 0377-2217,

Bath, S. K., Dhillon J. S. & Kothari, D. P. (2004), Fuzzy satisfying stochastic multi-objective generation scheduling by weightage pattern search methods, *Electric Power Systems Research*, Volume 69, Issues 2-3, Pages 311-320, ISSN

Bit, A. K., Biswal, M. P., & Alam, S. S. (1992), Fuzzy programming approach to multicriteria decision making transportation problem *Fuzzy Sets and Systems, 50*(2), 135-141.

Cardona-Valdés, Y., Álvarez, A. & Ozdemir, D., (2011) A bi-objective supply chain design problem with uncertainty, Transportation Research Part C: Emerging Technologies, Volume 19, Issue 5, Pages 821-832, ISSN 0968-090X

Carlsson, C. and Fullér, R., (2002), A position paper on the agenda for soft decision analysis, Fuzzy Sets and Systems, Volume 131, Issue 1, 1 Pages 3-11, ISSN 0165-0114

Chan, P., Hui, C., Li, W., Sakamoto, H. Hirata, K., & Li, P. (2006), Long-term electricity contract optimisation with demand uncertainties. *Energy, 31*(13), 2469-2485.

Chazal, M., Jouini, E. & Tahraoui, R (2007), Production planning and inventories optimisation: A backward approach in the convex storage case, *Journal of Mathematical Economics*, 44 997-1023

Chen, CL. & Lee, WC, (2004), Multi-objective optimisation of multi-echelon supply chain networks with uncertain demands and prices, *Computers & Chemical Engineering*, 28 1131-1144

Chen, C.L., Yuan, TW., & Lee, WC. (2007), Multi-criteria fuzzy optimisation for locating warehouses and distribution centers in a supply chain network. *Journal of the Chinese Institute of Chemical Engineers, 38*(5-6), 393-407.

Conner, B.S., (1967), The Manufacture of Nitric Acid. The role of platinum alloy gauzes in the ammonia oxidation process, *Platinum Metals Rev.*

Cooper W.W., Deng H., Huang Z. & Li S.X. (2004), Chance constrained programming approaches to congestion in stochastic data envelopment analysis, *European Journal of Operational Research*, Volume 155, Issue 2, Pages 487-501, ISSN 0377-2217

Deng, X., Li, Z., & Wang, S. (2005), A minimax portfolio selection strategy with equilibrium. *European Journal of Operational Research, 166*(1), 278-292.

Díaz, M. S., & Bandoni, J. A. (1996), A mixed integer optimisation strategy for a large scale chemical plant in operation. *Computers & Chemical Engineering, 20*(5), 531-545.

Dondo, R., Méndez, C. A., & Cerdá, J. (2008), Optimal management of logistic activities in multi-site environments. *Computers & Chemical Engineering, 32*(11), 2547-2569.

Dorneanu, B., Bildea, C. S., & Grievink, J. (2008), Model reduction techniques for dynamic optimisation of chemical plants operation. In Bertrand Braunschweig and Xavier Joulia (Ed.), *Computer aided chemical engineering; 18th European symposium on computer aided process engineering* (pp. 337-342) Elsevier.

Duh, J.D. and Brown, D.G.,(2007), Knowledge-informed Pareto simulated annealing for multi-objective spatial allocation, Computers, Environment and Urban Systems, Volume 31, Issue 3, Pages 253-281, ISSN 0198-9715

Elferchichi, A., Gharsallah, O., Nouiri, I., Lebdi, F., & Lamaddalena, N. (2009). The genetic algorithm approach for identifying the optimal operation of a multi-reservoirs on-demand irrigation system. *Biosystems Engineering, 102*(3), 334-344.

Ekel, P., Pedrycz, W. & Schinzinger, R., (1996), A general approach to solving a wide class of fuzzy optimisation problems, *Fuzzy Sets and Systems*, 97 49-66

Escudero, L.F., Galindo, E., Garcia, G., Gomez, E. & Sabau, V., (1998), Schumann, a modeling framework for supply chain management under uncertainty, *European Journal of Operational Research*, 119 14-34

Li, F., Meindersma, G.W., de Haan, A.B. & Reith, T., (2002), Optimisation of non-woven spacers by CFD and validation by experiments, *Desalination*, Volume 146, Issues 1-3, Pages 209-212

Feng, Y., D'Amours S. & Beauregard, R. (2007), The value of sales and operations planning in oriented strand board industry with make-to-order manufacturing system: Cross functional integration under deterministic demand and spot market recourse, *Int J Production Economics*, 925-5273

Ferrio, J., & Wassick, J. (2008), Chemical supply chain network optimisation. *Computers & Chemical Engineering, 32*(11), 2481-2504.

FSSA (2005), "Fertiliser use by crop in SA" *Food and Agriculture Organisation of the United Nations, Rome*, 23-30

Ghalebsaz-Jeddi, B., Shultes, B.C., & Haji, R. (2004), A multi-product continuous review inventory system with stochastic demand, backorders, and a budget constraint. *European Journal of Operational Research, 158*(2), 456-469.

Gibilisco, P. and Isola, T.(2011), On a refinement of Heisenberg uncertainty relation by means of quantum Fisher information, Journal of Mathematical Analysis and Applications, Volume 375, Issue 1, Pages 270-275, ISSN 0022-247X

Gjerdrum, J., Shah, N., & Papageorgiou, G.L. (2004), Fair transfer price and inventory holding policies in two-enterprise supply chains, *European Journal of Operational Research*, 143 582-999

Goicoechea, A & Duckstein, L. (1987), Non-normal deterministic equivalents and a transformation in stochastic mathematical programming, *Applied Mathematics and Computation*, Volume 21, Issue 1, Pages 51-72,

Grigori, D., Casati, F, Castellanos, D., Sayal, U. & Shan, M. (2003). Business Process Intelligence. *Computers in Industry, 53*, 321 - 343.

Grossmann, I.E. & You, F, (2008), Design of responsive supply chains under demand uncertainty, *Chemicals & Chemical Engineering*, 32(2008) 3090-3111

Guillen, G., Mele, F.D., Bagajewicz, M.J., Espuna, A. & Puigjaner, L., (2004), Multiobjective supply chain design under uncertainty, *Chemical Engineering Science*, 60 1535-1553

Guo P., Huang G.H. & Li Y.P. (2010), An inexact fuzzy-chance-constrained two-stage mixed-integer linear programming approach for flood diversion planning under multiple uncertainties, *Advances in Water Resources*, Volume 33, Issue 1, Pages 81-91

Gupta, A. & Maranas, C.D. (2003), Managing demand uncertainty in supply chain planning. *Computers & Chemical Engineering, 27*(8-9), 1219-1227.

Hannan, E.L. (1981), Linear programming with multiple fuzzy goals, (1981), Fuzzy Sets and Systems 6 (3) 235–248.

Haimes, Y.Y & Chankong, V. (1979), Kuhn-Tucker multipliers as trade-offs in multiobjective decision-making analysis, Automatica, Volume 15, Issue 1, January, Pages 59-72, ISSN 0005-1098

Ho, W., & Emrouznejad, A. (2009). Multi-criteria logistics distribution network design using SAS/OR. *Expert Systems with Applications, 36*(3, Part 2), 7288-7298.

Jackson, R. (1964). Some algebraic properties of optimisation problems in complex chemical plants. *Chemical Engineering Science, 19*(1), 19-31.

Janak, S.L., Lin, J.X., Christodoulos A. & Floudas, (2007), A new robust optimisation approach for scheduling under uncertainty: II.

Uncertainty with known probability distribution, *Computers & Chemical Engineering, Volume 31, Issue 3, 171-195*

Jiang L, Adam H., Nicholas P, Jianhua Y., & Summers R.M. (2007), "Validating Pareto optimal operation parameters of polyp detection algorithms for CT colonography", Proc. SPIE 6514, 65142G

Jung, J.Y., Blau, G., Pekny, J.F., Reklaitis, G. V., & Eversdyk, D. (2004). A simulation based optimisation approach to supply chain management under demand uncertainty. *Computers & Chemical Engineering, 28*(10), 2087-2106.

Kok, N, Verville, J, Danesh-Pajou, A & DeLuca, D. (2008). Communication flow orientation in business process modeling and its effect on redesign success: Results from a field study. *Decision Support Systems.*

Lee S.M., & Olson D.L.(1985), A gradient algorithm for chance constrained non-linear goal programming, *European Journal of Operational Research,* Volume 22, Issue 3, 5, Pages 359-369, ISSN 0377-2217,

Li, C., He, X., Chen, B., Xu, Q., & Liu, C. (2008), A hybrid programming model for optimal production planning under demand uncertainty in refinery. Chinese Journal of Chemical Engineering, *16*(2), 241-246.

Liang R & Jinwu G (2008), Dependent-Chance Programming Models for Capital Budgeting in Fuzzy Environments, Tsinghua Science & Technology, Volume 13, Issue 08, Pages 117-120.

Lin, C., & Chen, Y.M. (2009), Hedging strategic flexibility in the distribution optimisation problem. Omega, *37*(4), 826-837.

Luhandjula, M.K., & Gupta, M.M., (1995), On fuzzy stochastic optimisation, *Fuzzy Sets and Systems,* 47-55

Lukasiewicz, T. & Straccia, U. (2009), Description logic programs under probabilistic uncertainty and fuzzy vagueness, International Journal of Approximate Reasoning, Volume 50, Issue 6, June 2009, Pages 837-853, ISSN 0888-613X

Kim, J.S., & Whang, K.S., (1998), A tolerance approach to the fuzzy goal programming problems with unbalanced triangular membership function, *European Journal of Operational Research* 107 (1998) 614–624.

Kunsch, P.L & Teghem J.J, (1987), nuclear fuel cycle optimisation using multi-objective stochastic linear programming, *European Journal of Operational Research, 31, 240 – 249*

Li, P., Arellano-Garcia, H. & Wozny, G., (2008), Chance constrained programming approach to process optimisation under uncertainty, *Computers & Chemical Engineering,* Volume 32, Issues 1-2, 25-45

Liu, S. & Papageorgiou, L.G. (2013), Multiobjective optimisation of production, distribution and capacity planning of global supply chains in the process industry, Omega, Volume 41, Issue 2, April 2013, Pages 369-382, ISSN 0305-0483, 10.1016

Mahapatra, G.S., & Roy, T.K., (2002), Fuzzy multi-objective mathematical programming on reliability optimisation model, *Applied Mathematics and Computation,* 127 141-164

Martin, C.H. & Lubin, S.L. (1985), Optimisation Modeling for Business Planning at Turnbull Asphalt. *Interfaces, 15,* 66-721.

Massimiliano A., & Chiandussi, G. (2007), Optimisation of a vehicle energy absorbing steel component with experimental validation, *International Journal of Impact Engineering,* Volume 34, Issue 407, Pages 843-858

Mavrotas, G, & Florios, K, (2013), An improved version of the augmented ε-constraint method (AUGMECON2) for finding the exact pareto set in multi-objective integer programming problems, Applied Mathematics and Computation, Volume 219, Issue 18, Pages 9652-9669, ISSN 0096-

Mertins, K, Jochem, R, & Jakel, F.W. (1997), A tool for object-oriented modeling and analysis of business processes. *Computers in Industry, 33,* 345 - 356.

Mesfin G. & Shuhaimi M. (2010), A chance constrained approach for a gas processing plant with uncertain feed conditions, Computers & *Chemical Engineering,* Volume 34, Issue 8, Pages 1256-1267, ISSN 0098-1354

Nagurney, A. (2009). A system-optimisation perspective for supply chain network integration: The horizontal merger case. Transportation Research Part E: Logistics and Transportation Revie*w, 45*(1), 1-15.

Narasimhan, R, (1980), Goal programming in a fuzzy environment, *Decision Sciences* 11 325–336.

Novak, D.C., Ragsdale, C.T., (2002), A decision support procedure for stochastic multi-criteria linear programming using spreadsheets, *Decision Support Systems,* Vol. P 36, 99-116

Olanipekun, E.O. (2003). Kinetics of acidulation reaction in nitrophosphate process. *International Journal of Mineral Processing, 69*(1–4), 1-9.

Punnen, A.P. & Aneja, Y.P., (1993), Minmax combinatorial optimisation, *European Journal of Operational Research,* 81 634-643

Raj, S. & Lakshminarayanan, T., (2008), Multi-objective Optimisation in Multi-echelon Decentralized Supply Chains, *Industrial Engineering Chemical Resources,* 47, 6661-66710

Rong, A. & Lahdelma, R. (2008), Fuzzy chance constrained linear programming model for optimizing the scrap charge in steel production, *European Journal of Operational Research,* Volume 186, Issue 3, Pages 953-964, ISSN 0377-2217

Rong, M., Mahapatra, N. K., & Maiti, M. (2008). A multi-objective wholesaler-retailers inventory-distribution model with controllable lead-time based on probabilistic fuzzy set and triangular fuzzy number. *Applied Mathematical Modelling, 32*(12), 2670-2685.

Roghanian, E., Sadjadi, S.J. & Aryanezhad, M.B. (2007), A probabilistic bi-level linear multi-objective programming problem to supply chain planning, *Applied Mathematics and Computation*, Volume 188, Issue 1, Pages 786-800

Sakawa, M., Yano, H & Takahashi, J, (1992), Pareto optimality for multiobjective linear fractional programming problems with fuzzy parameters, Information Sciences, Volume 63, Issues 1–2, Pages 33-53, ISSN 0020-0255,

Santoso, T, Ahmed, S, Goetschalckx, M, & Alexander, A, (2004), A stochastic programming approach for supply chain network design under uncertainty, Georgia Institute of Technology, USA

Sheu, J. (2007), A hybrid fuzzy-optimisation approach to customer grouping-based logistics distribution operations. *Applied Mathematical Modelling, 31*(6), 1048-1066.

Smith, J.R. & Smith, J.W., (2006), A relationship between the carbon and hydrogen content of coals and their vitrinite reflectance, *Coal Geology*, 70 79-86

Statnikov, R., Bordetsky, A., Matusovb, A., Il'ya Sobol, J., & Statnikov, A., (2008), Definition of the feasible solution set in multi-criteria optimisation problems with continuous, discrete, and mixed design variables, *Non-linear Analysis*,

Tamiz, M., Mirrazavi, S.K, & Jones, D.F. (1999), Extensions of Pareto efficiency analysis to integer goal programming, *Omega*, Volume 27, Issue 2, Pages 179-188, ISSN 0305-0483,

Tsai, J. (2007), An optimisation approach for supply chain management models with quantity discount policy. *European Journal of Operational Research, 177*(2), 982-994.

Tsiakis, P. &, Papageorgiou, L.G. (2008), Optimal production allocation and distribution supply chain networks, International Journal of Production Economics, Volume 111, Issue 2, *Special Section on Sustainable Supply Chain*, Pages 468-483, ISSN 0925-5273

Vergidis, K, Tiwan, A & Majeed, B. (2007), Business Process Improvement using multi-objective optimisation. *Technology Journal, 24, 2.*

Vergidis, K, Tiwan, A, Majeed, B & Roy, R. (2006). Optimisation of business process designs: An algorithmic approach with multiple objectives. *International Journal of Production Economics, 1096,* 105-121.

Wu, H.C., (2009), The Karush-Kuhn-Tucker optimality conditions in multiobjective programming problems with interval-valued objective functions, *European Journal of Operational Research,* Volume 196, Issue 1, Pages 49-60, ISSN

Xevi, E., & Khan, S. (2005). A multi-objective optimisation approach to water management. *Journal of Environmental Management, 77*(4), 269-277.

Xu, J, Liu, Q, & Wang, R. (2008), A class of multi-objective supply chain networks optimal model under random fuzzy environment and its application to the industry of Chinese liquor, *Information Sciences,* Volume 178, Issue 8, Pages 2022-2043, ISSN 0020-0255

Yaghoobi, M.A., & Tamiz, M. (2007). A method for solving fuzzy goal programming problems based on MINMAX approach. *European Journal of Operational Research, 177*(3), 1580-1590.

Yang, G, Wang, Z & Li, X, (2008), The optimisation of the closed-loop supply chain network, *Transportation Research Part E,* 45 16-28

Mavrotas, G, & Florios, K, An improved version of the augmented ε-constraint method (AUGMECON2) for finding the exact pareto set in multi-objective integer programming problems, Applied Mathematics and Computation, Volume 219, Issue 18, 15 May 2013, Pages 9652-9669, ISSN 0096-

Zhang, J., Chen, J., & Lee, C. (2008). Joint optimisation on pricing, promotion and inventory control with stochastic demand. *International Journal of Production Economics, 116*(2), 190-198.

Zhu, X., Ma, Q., Zhu, Z, & Cheng, J. (2006), Validation of an optimisation procedure to improve low frequency characteristics of rooms, *Applied Acoustics,* Volume 67, Issue 6, Pages 529-540

Zimmerman, H.J. (1978), 'Fuzzy programming and linear programming with several objective functions', *Fuzzy Sets and Systems,* (1) 45 – 55

INDEX

www.ingramcontent.com/pod-product-compliance
Ingram Content Group UK Ltd.
Pitfield, Milton Keynes, MK11 3LW, UK
UKHW021033041026
15470UKWH00017B/1834/J

9 781622 730162